Low Voltage
Electron Microscopy

Current and future titles in the Royal Microscopical Society—John Wiley Series

---

Published

Principles and Practice of Variable Pressure/Environmental Scanning Electron Microscopy (VP-ESEM)
Debbie Stokes

Aberration-Corrected Analytical Electron Microscopy
Edited by Rik Brydson

Diagnostic Electron Microscopy—A Practical Guide to Interpretation and Technique
Edited by John W. Stirling, Alan Curry & Brian Eyden

Low Voltage Electron Microscopy—Principles and Applications
Edited by David C. Bell & Natasha Erdman

Forthcoming

Atlas of Images and Spectra for Electron Microscopists
Edited by Ursel Bangert

Understanding Practical Light Microscopy
Jeremy Sanderson

Focused Ion Beam Instrumentation: Techniques and Applications
Dudley Finch & Alexander Buxbaum

Electron Beam-Specimen Interactions and Applications in Microscopy
Budhika Mendis

Low Voltage Electron Microscopy: Principles and Applications

Edited by

David C. Bell
Harvard University, USA

and

Natasha Erdman
JEOL USA Inc., USA

Published in association with the Royal Microscopical Society

Series Editor: Susan Brooks

RMS

WILEY
A John Wiley & Sons, Ltd., Publication

This edition first published 2013

Registered office
John Wiley & Sons Ltd, The Atrium, Southern Gate, Chichester, West Sussex, PO19 8SQ, United Kingdom

For details of our global editorial offices, for customer services and for information about how to apply for permission to reuse the copyright material in this book please see our website at www.wiley.com.

Library of Congress Cataloging-in-Publication Data

Low voltage electron microscopy : principles and applications / edited by David C. Bell, Natasha Erdman.
pages cm.
Includes index.
ISBN 978-1-119-97111-5 (hardback)
1. Electron microscopy–Technique. I. Bell, D. C. (David C.) II. Erdman, Natasha.
QH212.E4L69 2012
502.8′25–dc23

2012033919

A catalogue record for this book is available from the British Library.

HB ISBN: 978-1-119-97111-5

Set in 10.5/13pt Sabon by Laserwords Private Limited, Chennai, India

Contents

List of Contributors ix
Preface xi

1 Introduction to the Theory and Advantages of Low Voltage Electron Microscopy 1
David C. Bell and Natasha Erdman
1.1 Introduction 1
1.2 Historical Perspective 2
1.3 Beam Interaction with Specimen—Elastic and Inelastic Scattering 3
1.3.1 The Scattering Cross Section 6
1.3.2 Effects of Specimen Damage 10
1.4 Instrument Configuration 11
1.4.1 Scanning Electron Microscope 11
1.4.2 Transmission Electron Microscope 12
1.4.3 Scanning Transmission Electron Microscope 12
1.5 Influence of Electron Optics Aberrations at Low Voltages 12
1.5.1 Spherical Aberration 13
1.5.2 Effect of Chromatic Aberration 14
1.5.3 The Diffraction Limit 15
1.5.4 Optimizing Spot Size for SEM and STEM 15
1.6 SEM Imaging at Low Voltages 16
1.6.1 Primary Contrast Signals and their Detection in SEM 18
1.6.2 Backscattered Electrons 18
1.6.3 Secondary Electrons 21
1.6.4 Charge Balance in SEM 23

1.6.5 SEM Image Contrast 24
1.6.6 Microanalysis in SEM at Low Voltages 25
1.7 TEM/STEM Imaging and Analysis at Low Voltages 26
1.8 Conclusion 27
References 28

2 SEM Instrumentation Developments for Low kV Imaging and Microanalysis 31
Natasha Erdman and David C. Bell
2.1 Introduction 31
2.2 The Electron Source 33
2.3 SEM Column Design Considerations 36
2.4 Beam Deceleration 41
2.5 Novel Detector Options and Energy Filters 43
2.5.1 Secondary Detectors 43
2.5.2 Backscatter Detectors 45
2.6 Low Voltage STEM in SEM 48
2.7 Aberration Correction in SEM 50
2.8 Conclusions 53
References 53

3 Extreme High-Resolution (XHR) SEM Using a Beam Monochromator 57
Richard J. Young, Gerard N.A. van Veen, Alexander Henstra and Lubomir Tuma
3.1 Introduction 57
3.2 Limitations in Low Voltage SEM Performance 58
3.2.1 Aberration Correction 58
3.2.2 Electron Source Energy Spread 59
3.3 Beam Monochromator Design and Implementation 59
3.4 XHR Systems and Applications 63
3.4.1 Elstar XHR Electron Column 64
3.4.2 Beam Deceleration for Extending Low-Voltage Performance 65
3.4.3 Combination of a Monochromator with Non-Immersion Lens 67
3.4.4 XHR Applications 68
3.5 Conclusions 69
Acknowledgements 70
References 70

4 The Application of Low-Voltage SEM—From Nanotechnology to Biological Research 73

Natasha Erdman and David C. Bell

4.1 Introduction 73
4.2 Specimen Preparation Considerations 74
4.3 Nanomaterials Applications 76
4.3.1 Nanoparticles, Nanotubes and Nanowires 76
4.3.2 Nanoporous Materials 81
4.3.3 Graphene 83
4.4 Beam Sensitive Materials 84
4.5 Semiconductor Materials 85
4.6 Biological Specimens 87
4.7 Low-Voltage Microanalysis 91
4.8 Conclusions 92
References 93

5 Low Voltage High-Resolution Transmission Electron Microscopy 97

David C. Bell

5.1 Introduction 97
5.2 So How Low is Low? 99
5.3 The Effect of Chromatic Aberration and Chromatic Aberration Correction 100
5.4 The Electron Monochromator 103
5.5 Theoretical Tradeoffs of Low kV Imaging 105
5.6 Our Experience at 40 keV LV-HREM 109
5.7 Examples of LV-HREM Imaging 110
5.8 Conclusions 114
References 116

6 Gentle STEM of Single Atoms: Low keV Imaging and Analysis at Ultimate Detection Limits 119

Ondrej L. Krivanek, Wu Zhou, Matthew F. Chisholm, Juan Carlos Idrobo, Tracy C. Lovejoy, Quentin M. Ramasse and Niklas Dellby

6.1 Introduction 119
6.2 Optimizing STEM Resolution and Probe Current at Low Primary Energies 121
6.3 STEM Image Formation 128
6.3.1 Basic Principles 128

6.3.2 ADF Imaging 132
6.4 Gentle STEM Applications 135
6.4.1 Single Atom Imaging 135
6.4.2 Single Atom Spectroscopy 146
6.4.3 Single Atom Fine Structure EELS 152
6.5 Discussion 154
6.6 Conclusion 156
Acknowledgements 157
References 157

7 Low Voltage Scanning Transmission Electron Microscopy of Oxide Interfaces 163
Robert Klie
7.1 Introduction 163
7.2 Methods and Instrumentation 166
7.3 Low Voltage Imaging and Spectroscopy 168
7.3.1 $SrTiO_3/BiFeO_3$ Interface 168
7.3.2 Si_3N_4/SiO_2 Interfaces 170
7.3.3 Ultrathin $SrTiO_3$ films on GaAs 175
7.4 Summary 180
Acknowledgements 180
References 180

8 What's Next? The Future Directions in Low Voltage Electron Microscopy 185
David C. Bell and Natasha Erdman
8.1 Introduction 185
8.2 Unique Low Voltage SEM and TEM Instruments 186
8.2.1 Miniature SEM Columns 186
8.2.2 Dedicated Low Voltage TEM 187
8.2.3 The Helium Ion Microscope as an Alternative to Low Voltage SEM Imaging 189
8.3 Cameras, Detectors, and Other Accessories 192
8.3.1 The Direct Electron Detector 192
8.3.2 Silicon Drift Detectors for Low kV Nanoanalysis 195
8.4 Conclusions 198
References 199

Index 201

List of Contributors

David C. Bell, School of Engineering and Applied Sciences, Harvard University, USA

Matthew F. Chisholm, Materials Science and Technology Division, Oak Ridge National Laboratory, USA

Niklas Dellby, Nion Co., USA

Natasha Erdman, JEOL USA Inc., USA

Alexander Henstra, FEI Company, The Netherlands

Juan Carlos Idrobo, Materials Science and Technology Division, Oak Ridge National Laboratory, USA

Robert Klie, Nanoscale Physics Group, Department of Physics, University of Illinois at Chicago, USA

Ondrej L. Krivanek, Nion Co., USA

Tracy C. Lovejoy, Nion Co., USA

Quentin M. Ramasse, SuperSTEM Laboratory, STFC Daresbury, UK

Lubomir Tuma, FEI Company, Czech Republic

Gerard N.A. van Veen, FEI Company, The Netherlands

Richard J. Young, FEI Company, USA

Wu Zhou, Department of Physics & Astronomy, Vanderbilt University, USA and Materials Science and Technology Division, Oak Ridge National Laboratory, USA

Preface

The introduction of the electron microscope in the 1930s enabled the visual confirmation that virus particles actually existed for the first time and were separate individual "living" entities. The new technology advances of electron microscopes at that time period, in particular those manufactured by Siemens in Germany and the Radio Corporation of America (RCA) enabled the routine imaging of biological tissues and cells. Further advances allowed the detailed identification and characterization of viruses, as better techniques and more powerful microscopes were developed. Electron microscopic imaging was now in a dramatic new way completely beyond the capabilities of light microscopy.

The electron microscope was invented by Max Knoll and Ernst Ruska in 1931, although there were many researchers working on the development of electron microscopes. Ernst Ruska was awarded half of the Nobel Prize for Physics in 1986 for his invention. (The other half of the Nobel Prize was divided between Heinrich Rohrer and Gerd Binnig for the STM.)

Over the years electron microscopes became more powerful in terms of resolution and by the 1970s atomic structure imaging was routinely possible. By the end of the 1990s sub-Ångstrom resolution was realized following on with the development of commercial aberration correctors. Higher power in terms of resolution has been the typical benchmark by which to measure the development of the electron microscope both for the SEM and the TEM; however, with new instrumental advances it now becomes possible for greatly reduced voltage instrument to not only provide the same resolution but rather also improved contrast at the same time.

There are currently multiple texts that address either the general topics in electron microscopy (theory, operation and applications of SEM, TEM or STEM) or more specific subjects, like FIB, ESEM, bio- or polymer microscopy. The intent of this volume is to discuss the rapidly developing and cutting edge field of low voltage microscopy that has only recently become available due to the rapid developments in the electron optics design and image processing. The low voltage techniques are particularly crucial for nanotechnology and research of the surface related phenomena, allowing researches to observe materials as has never been done before. The most recent volume by Scatten and Pawley (eds) published in 2008 focuses solely on the SEM aspects of low voltage observation of biological materials. The volume presented here addresses the recent developments in theory and instrumentation and serves as a practical guide to the current and new microscopists and materials scientists who are active in the field of nanotechnology and biological imaging. The premise of this book is not to cover every single subject under the umbrella of the "low voltage microscopy" but to target principles and practical aspects of this emerging field as it pertains to the imaging techniques in widely used commercial instruments, such as SEM, TEM and STEM.

The opening chapter discusses the general concepts and principles of specimen interaction with electron beam, electron microscope instrumentation and low voltage imaging in SEM and TEM/STEM. The following Chapters (2–5) focus on SEM instrumentation and applications. Chapter 2 (N. Erdman and D.C. Bell) attempts to capture recent developments in SEM instrumentation as they pertain to low voltage microscopy and microanalysis. Chapter 3 is dedicated to development and use of monochromator in SEM (R. Young *et al.*). Chapter 4 discusses practical aspects of sample preparation, imaging and microanalysis for nanostructured and beam sensitive materials using low voltage. This chapter also references and shows examples of biological imaging with low voltages; however, since there exists a recent volume on this very topic, our goal is to present additional examples.

Chapters 5–7 talk about low voltage applications in TEM and STEM. Chapter 6 by D.C. Bell shows use of aberration corrected and monochromated TEM for imaging of beam sensitive materials down to 40 kV. Krivanek *et al.* show utilization of the "Gentle STEM" technique in Chapter 6 and R.F. Klie discusses STEM imaging and microanalysis of oxide interfaces in Chapter 7.

In the final chapter we have dedicated some discussion to other emerging techniques for low voltage imaging and analysis, such as miniature SEM columns, dedicated low voltage TEM instrumentation as well as utilization of Helium ion microscopy as an alternative to low voltage scanning electron microscopy imaging. Our hope is that current and aspiring electron microscopists, nanotechnology, materials science and biology researchers who routinely use electron microscopes for their research will benefit from the theoretical and practical discussions provided in this volume.

We would like to extend our sincere thanks to all the contributors. N. Erdman would like to extend special thanks to A. Laudate of JEOL USA for fruitful discussions and providing some of the excellent graphics used in this volume. The editors would like to thank A. Lazar of Carl Zeiss Microimaging for providing us with relevant micrographs and illustrations. We would also like to thank Professor David Joy for his invaluable insight and advice.

David C. Bell, Cambridge, MA
Natasha Erdman, Peabody, MA

1

Introduction to the Theory and Advantages of Low Voltage Electron Microscopy

David C. Bell[1] **and Natasha Erdman**[2]
[1]*School of Engineering and Applied Sciences, Harvard University, USA*
[2]*JEOL USA Inc., USA*

1.1 INTRODUCTION

The fundamental aspects of electron microscopy all relate directly to the physics of the interactions between the electron beam and sample. These interactions have been studied extensively since the discovery of the electron by J.J. Thompson in 1897. Energetic electrons are described as "ionizing radiation"—the general term used to describe radiation that is able to ionize or remove the tightly bound inner shell electrons from a material. This is obviously an advantage for electron microscopy in that it produces a wide range of secondary signals such as secondary electrons and X-rays, but is also a disadvantage from the perspective that the sample is "ionized" by the electron beam and possibly structurally damaged, which depending on the accelerating voltage happens in a number of different ways. The advantages of using a lower accelerating voltage for the electron beam are that the energy is reduced and hence the momentum that *can* be transferred to sample from the electron is also reduced. This, however, has the unwanted effect of reducing the possible emitted signal; although, with recent improvements in detectors,

Low Voltage Electron Microscopy: Principles and Applications, First Edition.
Edited by David C. Bell and Natasha Erdman.

cameras and the use of aberration correctors, the signal to noise and the resolution to produce a final image can not only be maintained but are actually improved.

This chapter will detail the basic theory of electron beam interactions and how it relates to electron microscopy at low voltage. There are, however, distinct differences between the important considerations for low voltage SEM imaging as compared to TEM imaging and these will be detailed in the text.

1.2 HISTORICAL PERSPECTIVE

The early steps in the development of the electron microscope in the 1930s and 1940s by different research groups led ultimately to the development of two distinct groups of instruments: the scanning electron microscope (SEM) and the transmission (or scanning transmission) electron microscope (TEM and STEM). The early microscope designs by Knoll and by Ruska (1933) showed transmission electron images of solid surfaces at 10–16X magnification, which was improved upon by introduction of replica sample preparation technique for TEM observation (Mahl, 1940). As a continuation of his work with Ruska, M. Knoll had designed an electron beam scanner in 1935 (Knoll, 1935) to study targets for the TV camera tubes; this was in essence a predecessor to an SEM, with accelerating voltage up to 4 kV. In 1936 through his contract with the company Siemens, Manfred von Ardenne began development of a scanning transmission electron microscope, mainly to avoid detrimental effects of chromatic aberration during observation of thick specimens in TEM. The microscope built by von Ardenne had a probe size of 4 nm (von Ardenne, 1937; von Ardenne 1938). The work by von Ardenne, though interrupted by the events of World War II, nonetheless established a theoretical and design background for future SEM and STEM development, particularly regarding understanding of beam/specimen interactions, effect of accelerating voltage on resolution, as well as detector design and positioning within the microscope (von Ardenne, 1985).

From 1938 to 1942, V. Zworykin at RCA headed parallel SEM and TEM development projects that resulted in an SEM instrument with accelerating voltage of 800V (Zworykin *et al.*, 1942). However, poor vacuum in the system significantly impacted the resulting micrographs, and the quality of the recorded images was disappointing with mostly topographic contrast and no meaningful compositional information. These results prompted RCA to discontinue the SEM project and concentrate on the development of the TEM instrument, and led to the development of several commercial instruments. Nonetheless, the work on SEM instrument development continued in Cambridge in early 1950s (McMullan, 1952; McMullan, 2004). R.F.M. Thornley successfully developed the first low voltage SEM (Thornley, 1960) in Oatley's lab at Cambridge University in the early 1960s. By improving upon the existing SEM2 design (Wells, 1957), he was able to obtain 200 nm probe at 1 kV. Prior to that experiment, the SEM was always operated at higher voltages (greater than 6 kV) that allowed only observation of conductive specimens. Thornley's work showed that a surface of alumina ceramic could be imaged at 1.5 kV negating charging artifacts (Thornley, 1960); moreover, he recognized the importance of low voltage in reducing the charge build up that had caused issues in non-conductive samples.

Over the years, significant improvements in electronics, vacuum and electron column design, as well as detector technology have improved SEM instrument performance to the level where the resolution at 1 kV is on the order of 1–2 nm for high-end field emission systems (see Chapter 2). The recent developments in aberration correction and addition of monochromators to TEM and STEM instruments have further improved their performance for both high and low accelerating voltage applications (see Chapters 6–8).

1.3 BEAM INTERACTION WITH SPECIMEN—ELASTIC AND INELASTIC SCATTERING

Interaction of a primary electron beam with specimen can generate several different signals (Figure 1.1)—secondary and backscatter electrons, transmitted electrons (if the specimen is sufficiently thin), Auger electrons, characteristic X-rays and photons.

The basic elastic and inelastic scattering processes and electron excitation in materials have a direct influence on the electron range and depth of ionization distribution as well as secondary and backscatter electron emission and the observed contrast in all types of electron microscopes. Particle model of elastic and inelastic scattering processes (based on Bohr atom model) is shown in Figure 1.2(a), while Figure 1.2(b) displays band structure with inelastic processes as well as Auger and X-ray emissions, with respect to different energy levels. Multiple elastic scattering events produce electron backscattering; additional multiple inelastic scattering processes lead to eventual energy loss along the electron trajectories deeper within the material that result in the electrons

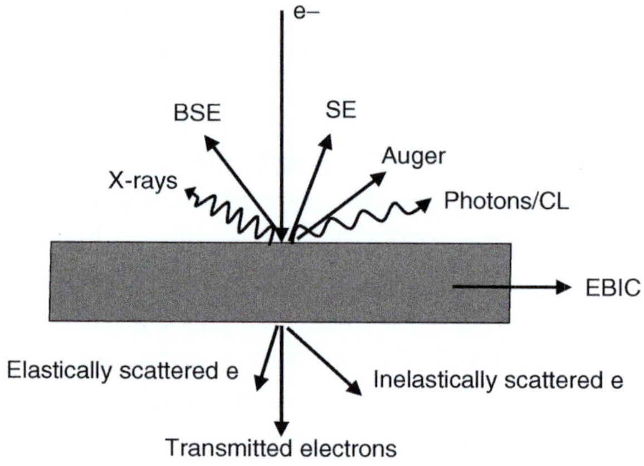

Figure 1.1 Overview of the signals generated when an electron beam interacts with a (relatively) thin specimen. In the case of a thick specimen there are no transmitted electrons and the signal gets absorbed within the material.

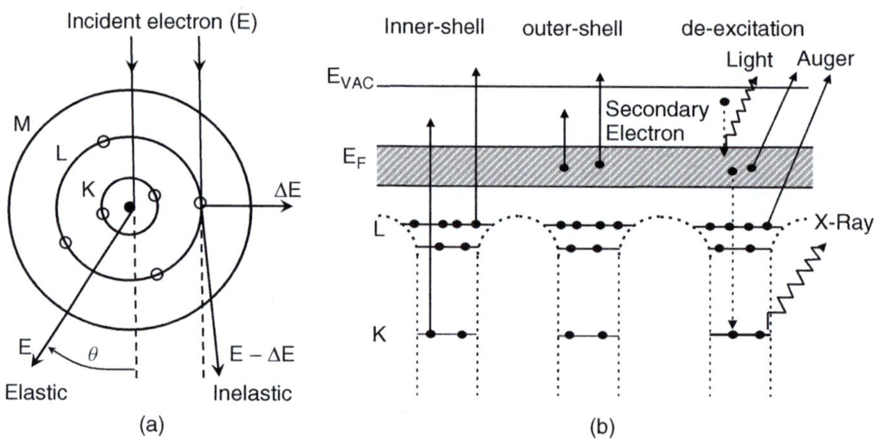

Figure 1.2 (a) Schematic of elastic and inelastic scattering due to the interaction of electron beam of energy E with an atom. (b) Diagram of inelastic excitations, X-ray, photon and Auger emissions with respect to different energy levels.

slowing down and eventually coming to rest. Inelastic scattering is also responsible for the generation of secondary electron signal, Auger electrons, X-rays, electron–hole pairs (semiconductors and insulators), cathodoluminescence and phonon and plasmon production. At lower accelerating voltages, the number of inelastic scattering events decreases;

for example, in Si K-shell ionization is no longer possible if accelerating voltage is below 1.84 kV, an effect known as the Duane-Hunt limit.

Understanding of the elastic and inelastic scattering processes can additionally serve as a basis for modeling of beam/specimen interactions (particularly for SEM imaging and analysis) via Monte–Carlo simulations to investigate electron trajectories in materials and calculate theoretical secondary and backscatter electron spatial distributions based on the specimen position under the beam (angle), accelerating voltage and the material type. Several different programs are available for these types of calculations; more specifically Casino (Drouin *et al.*, 2007; also http://www.gel.usherbrooke.ca/casino/index.html) has been written particularly with a focus on low voltage imaging and analysis.

The quantum mechanical properties of electron are such that the electron has a wavelength defined by de Broglie relationship:

$$\lambda = \frac{h}{p} = \frac{h}{\sqrt{2m_0E}} \quad (1.1)$$

where h is the Planck constant, m_0 is the rest mass of the electron and E the accelerating voltage. However, since electrons in the electron microscope are moving at high speeds defined by the accelerating voltage of the electron gun this equation needs to be rewritten be become the relativistic version (where E_0 is the rest energy):

$$\lambda = \frac{hc}{\sqrt{2EE_0 + E^2}} \quad (1.2)$$

This means that for a 200 kV electron we have a wavelength of 2.5 pm, but for a 40 kV electron the wavelength becomes nearly three times as large at 6 pm. If we plot wavelength as a function of accelerating voltage, we see that the wavelength increases exponentially as the accelerating voltage decreases below about 20 kV (Figure 1.3). For an SEM operating at 1 kV the wavelength becomes 39 pm, still a small value but a significantly larger wavelength as compared to, for example, 40 kV. The limiting factor for resolution of electron microscope is ultimately the electron wavelength; moreover, the actual working resolution of electron microscopes is directly limited by the lens aberrations present, as detailed later in this chapter. As Figure 1.1 shows, there is a variety of different possible signals generated by the interaction of primary beam with the specimen and to start with we separate these processes based on the differences in the scattering cross section between elastic and inelastic scattering.

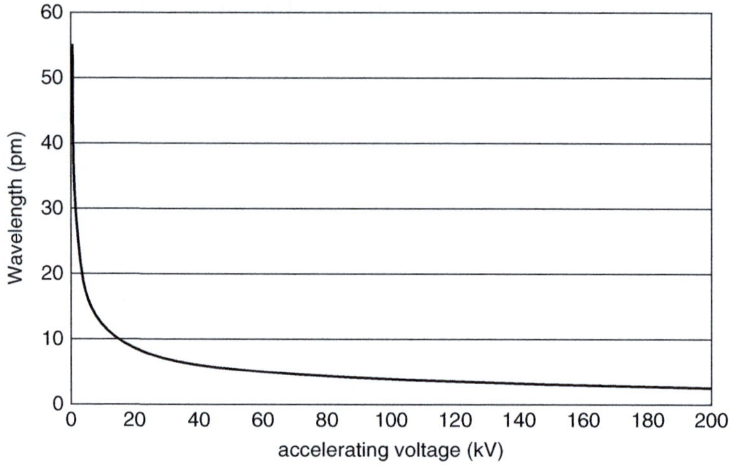

Figure 1.3 Dependence of electron wavelength on the accelerating voltage.

1.3.1 *The Scattering Cross Section*

The probability of an incident electron being scattered by a given atom per unit solid angle Ω is represented by the differential scattering cross section $d\sigma/d\Omega$ that is a function of the scattering angle θ (Figure 1.4). Interaction of the primary beam electrons with the attractive nucleus

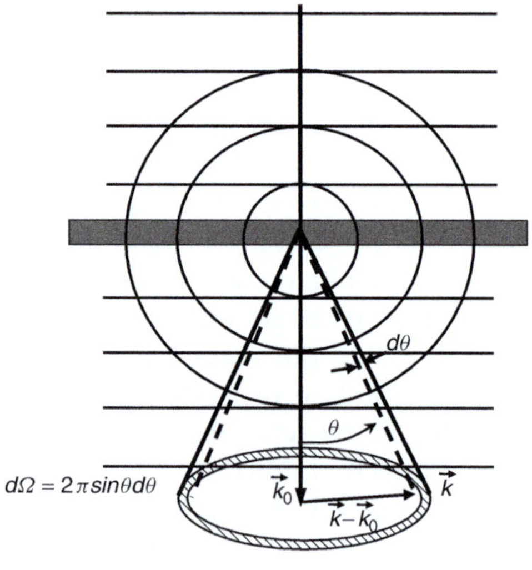

Figure 1.4 Electron scattering as a function of scattering angle θ. Superposition of an incident plane wave with amplitude $\psi = \psi_0 \exp(2\pi i k_0 z)$ and a spherical scattered wave with amplitude $\psi_{sc} = \psi_0 f(\theta) \exp(2\pi i k_0 r)/r$.

Coulomb potential results in elastic scattering processes, with change in a primary electron beam trajectories through angle θ, but negligible energy loss (on the order of meV). For elastic scattering, the cross section is given by:

$$\frac{d\sigma}{d\Omega} = |f(\theta)|^2 \tag{1.3}$$

where f is the scattering factor, which is a function of the scattering angle θ. Assuming a single scattering within each atom (the first Born approximation), scattering factor is proportional to the tree dimensional Fourier transform of the atomic potential $V(r)$. The elastic scattering cross section can also be expressed as a function of an elastic scattering factor $F(q)$:

$$\frac{d\sigma}{d\Omega} = \frac{4}{a_0^2 q^4}|F(q)|^2 = \frac{4\gamma^2}{a_0^2 q^4}\left|Z - f_x(q)\right|^2 \tag{1.4}$$

where $f_x(q)$ is the atomic scattering factor of an incident photon and equals the Fourier transform of the electron density within the atom; $a_0 = 4\pi\varepsilon_0\hbar^2/m_0e^2 = 0.529 \times 10^{-10}$ m is the Bohr radius, $\gamma = (1-v^2/c^2)^{-1/2}$ is a relativistic factor and $q = 2k_0\sin(\theta/2)$ is the scattering vector. The incident electrons are scattered by the entire electrostatic field of the atom (as opposed to X-rays that interact only with the atomic electrons), hence the inclusion of the atomic number (or nuclear charge) Z in Equation (1.4).

The relatively simple Rutherford scattering model (Rutherford, 1911) has been widely used in the past to describe elastic scattering of charged particles based on the unscreened electrostatic field of a nucleus. Setting the electronic term, $f_x(q)$, in Equation (1.4) to zero, the differential cross section becomes:

$$\frac{d\sigma}{d\Omega} = \frac{4}{a_0^2 q^4} Z^2\gamma^2 \tag{1.5}$$

The above equation is a reasonable approximation for light elements at large scattering angles (and has been often employed for modeling of high kV SEM imaging); however, at small scattering angles the equation breaks down, mainly due to the fact that the nuclear screening is not taken into account.

The nuclear screening can be incorporated via an expression for nuclear potential attenuated exponentially as a function of distance from the nucleus (r):

$$V(r) = \left(\frac{Ze}{4\pi\varepsilon_0 r}\right)\exp\left(-\frac{r}{r_0}\right) \tag{1.6}$$

where r_0 is the screening radius. The elastic scattering cross section then becomes:

$$\frac{d\sigma}{d\Omega} = \frac{4\gamma}{a_0^2}\left(\frac{Z}{q^4 + r_0^2}\right)^2 \approx \frac{4\gamma^2 Z^2}{a_0^2 k_0^4}\frac{1}{\left(\theta^2 + \theta_0^2\right)^2} \tag{1.7}$$

where $\theta_0 = (k_0 r_0)^{-1}$ and $r_0 = a_0 Z^{-1/3}$ (Lenz, 1954). The model in Equation (1.7) provides a quick estimate of the angular dependence of scattering; however, there are more sophisticated methods for calculating cross sections that take into account relativistic effects (for example Mott scattering cross section), and are particularly relevant at low incident energies.

A more accurate treatment of the elastic scattering cross section is achieved by taking into account relativistic quantum mechanics based on the Dirac equation. For example, Mott scattering cross-section (Mott, 1932; Mott and Massey, 1949; Reimer, 1993) incorporates the effect of spin-orbit coupling of electrons and thus differs from Rutherford scattering. Mott cross-section is especially applicable when describing beam-specimen interactions in the case of low voltage SEM observation. Mott cross-section is a quite complex function of primary electron energy, atomic number and scattering angle and several empirical models based on tabulated values have emerged and are being currently used in various beam/specimen interactions simulation programs (Czyzewski *et al.*, 1990; Gauvin and Drouin, 1993; Browning *et al.*, 1995; Hovington *et al.*, 1997).

Inelastic scattering is typically described as scattering events that result in energy loss of ΔE for the primary beam electrons through energy transfer to the atoms in the sample through interaction with either outer or inner shell atomic electrons; however, inelastic scattering does not significantly alter the electron trajectory.

The differential cross section for inelastic scattering can be written as (Reimer and Kohl, 2008):

$$\frac{d\sigma_i}{d\Omega} = \frac{4\gamma^2 Z}{a_0^2 q^4}\left(1 - \frac{1}{\left[1 + (qr_0)^2\right]^2}\right) \tag{1.8}$$

in which the scattering vector q is given by:

$$q^2 = k_0^2\left(\theta^2 + \overline{\theta}_E^2\right) \tag{1.9}$$

where $k_0 = 2\pi/\lambda$ is the magnitude of the incident electron wave vector, θ is the scattering angle, $\overline{\theta}_E = \overline{E}/\gamma m_0 v^2$ is the angle associated with the mean energy loss of $\overline{E}$. The above two expressions can be combined to give the expression for inelastic scattering cross section as a function of the scattering angle (Colliex and Mory, 1984):

$$\frac{d\sigma_i}{d\Omega} = \frac{4\gamma^2 Z}{a_0^2 k_0^4} \frac{1}{\left(\theta^2 + \overline{\theta}_E^2\right)^2} \left\{1 - \left[1 + \frac{\overline{\theta}_E^2}{\theta_0^2} + \frac{\theta^2}{\theta_0^2}\right]^{-2}\right\} \tag{1.10}$$

Figure 1.5 illustrates angular distribution of elastic and inelastic scattering cross sections calculated in carbon at an accelerating voltage of 100 kV showing how the scattering cross section decreases rapidly from zero degrees, interestingly of note is how the inelastic scattering

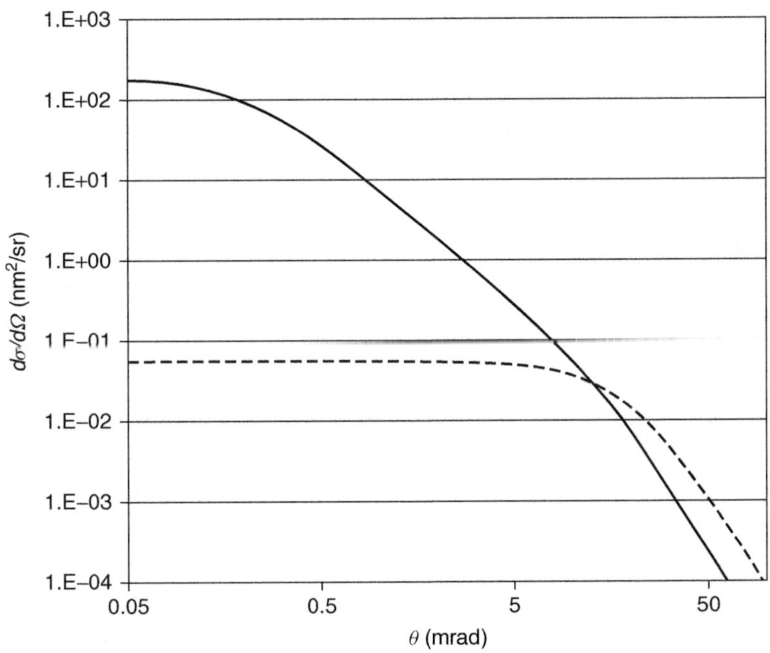

Figure 1.5 Angular dependence of elastic (dashed line) and inelastic (solid line) cross sections for carbon at accelerating voltage of 100 kV.

starts higher than elastic and then drops off more rapidly than the elastic. For more detailed discussion and analysis of differential cross sections see for example Egerton (2011) and Reimer and Kohl (2008).

1.3.2 *Effects of Specimen Damage*

Although elastic and inelastic scattering processes are essential for imaging and spectroscopy in electron microscope, they are also responsible for the electron beam induced specimen damage and alteration. The two most common damage mechanisms are termed radiolysis and knock-on damage.

Knock-on damage occurs when the incident electron energy is higher than the atomic sputtering threshold energy (Mott, 1932; Hobbs, 1979), thus inducing either an atom removal from its site or sputtering from the surface. The maximum energy that can be transferred to an atom during a collision is:

$$E_{\max} = \frac{2E(E + 2m_0c^2)}{M_0c^2} \tag{1.11}$$

where E in the incident electron energy, M_0 is the mass of the atom, and m_0c^2 is the rest energy of the electron. Equation (1.11) describes the threshold energy (E_{th}) needed for incident electrons to displace or sputter atoms.

Radiolysis damage is induced by converting the exciton energy generated by incident beam interaction with specimen into momentum and thus creating atomic displacements in the analyzed material (Hobbs and Pascucci, 1980; Pascucci, 1983). In order for radiolysis to occur the exciton energy should be larger than the energy needed for atomic displacement, and the exciton relaxation time should be sufficiently long ($\geq$1 ps) such that the bonding instabilities can be induced by mechanical relaxation of the atoms. Radiolysis processes are common in materials like organic compounds, silicates, halides and ice (Hobbs, 1979).

Some recent experimental observations and theoretical predictions suggest that below incident energies of 70 kV the damage is mainly radiolitic; whereas at incident energies above 200 kV the knock-on damage and material sputtering will be dominant (Ugurlu *et al.*, 2011). Therefore, TEM and STEM imaging at low voltages should substantially reduce knock-on damage.

1.4 INSTRUMENT CONFIGURATION

1.4.1 *Scanning Electron Microscope*

The scanning electron microscope (SEM) operates by rastering a fine electron probe (few nm in size) over a region of the specimen. Typical scanning electron microscope configuration is shown in Figure 1.6(a). Contemporary SEM instruments usually operate between few tens of volts and 30 kV, and are traditionally used for examination of bulk materials, though very thin specimens can be examined in transmission mode as well. The spatial resolution of the instrument is characterized by the electron probe diameter that can be achieved with the combination of electron source size and the lens configuration; however, the beam/specimen interaction volume further limits the instrument resolution. The instrument can collect (depending on the detector configuration) secondary, backscatter and transmitted electron signals. Moreover, signals like X-ray emission (EDS and WDS), cathodoluminescence (CL), electron backscatter diffraction (EBSD) and electron beam induced current (EBIC) can be detected.

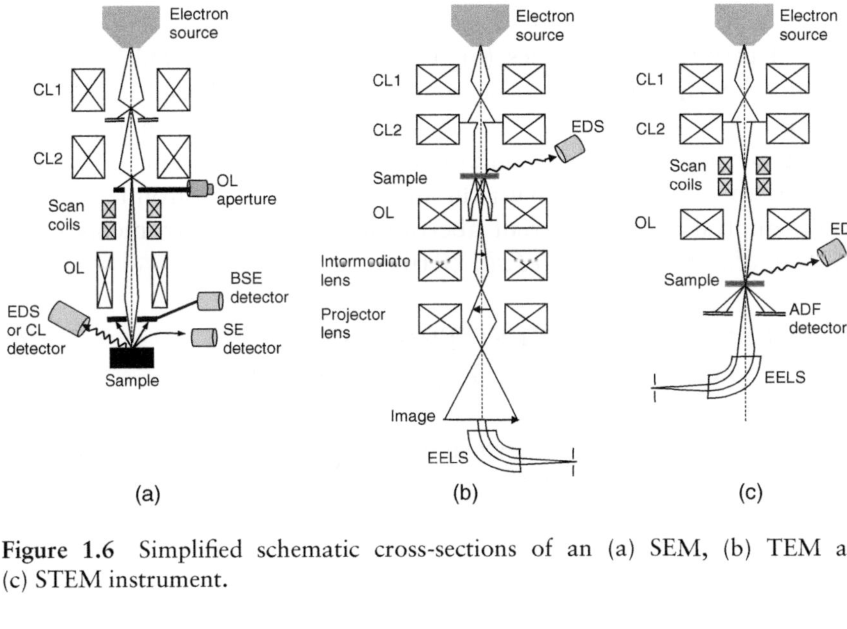

Figure 1.6 Simplified schematic cross-sections of an (a) SEM, (b) TEM and (c) STEM instrument.

1.4.2 *Transmission Electron Microscope*

In transmission electron microscope (TEM) a very thin specimen (less than 100 nm thick) is irradiated by a beam of high-energy electrons (between few tens of kV up to 1 MeV). A schematic of a traditional TEM configuration is presented in Figure 1.6(b). The TEM allows examination of specimens with an order of magnitude higher resolution than an SEM; the resolution of a non-aberration corrected TEM with field emission source is ~1.2 Å, whereas contemporary aberration corrected systems have been shown improve resolution down to 0.5 Å. Crystallographic information can be obtained in TEM using either selected area or convergent beam diffraction modes; elemental characterization can be obtained by utilizing either energy dispersive spectroscopy (EDS) or electron energy loss spectroscopy (EELS).

1.4.3 *Scanning Transmission Electron Microscope*

A dedicated scanning transmission electron microscope (STEM) utilizes a field emission gun that generates a probe size on the order of 1 Å that is rastered across an electron transparent specimen (Figure 1.6c). The instrument is typically operated at 80–300 kV accelerating voltage. The signal detection relies on the angular distribution of the scattered electrons with the ability to detect both electrons that are elastically scattered through high angles (high annular dark field, HAADF) and electrons that are inelastically scattered through small angles. HAADF signal contrast is indicative of the atomic number distribution (also called Z-contrast). The addition of EDS and EELS spectrometers further enhances the microscope, with the ability to perform atomic column specific analysis on the current aberration corrected instruments (e.g., Muller *et al.*, 2008; Botton *et al.*, 2010).

1.5 INFLUENCE OF ELECTRON OPTICS ABERRATIONS AT LOW VOLTAGES

The electron optical system performance is most often characterized by the minimum size of the electron probe that can be produced and the corresponding beam current in that probe size. However, the existing lens aberrations in the electron optical system assure that there is a limit to how small the spot of the electron beam can be focused into and the probe current that can be delivered in that spot. From the brightness equation:

$$\beta = \frac{I_b}{\pi.d_g^2/4}\,\frac{1}{\pi.\alpha^2} = \frac{4I_b}{\pi^2.\alpha^2.d_g^2} \tag{1.12}$$

it is clear, that the amount of current (I_b) for a given Gaussian spot size d_g will depend on the square of the convergence angle (α^2), while the brightness (β) is constant.

The lens aberrations are often classified by their order (Hawkes and Kasper, 1985)—defocus (first order), chromatic aberration (second order) and spherical aberration (third order). There are also first and third order astigmatism as well as axial coma (second order), which can be corrected by a more precise mechanical and electromagnetic column alignment, as well as use of smaller beam aperture sizes.

1.5.1 *Spherical Aberration*

Figure 1.7(a) shows a schematic representation of cause of spherical aberration in a round lens. The parallel rays incident on the electron lens are focused at different planes based on their angle of incidence and the refractive index of the lens; the rays traveling at high angles are focused

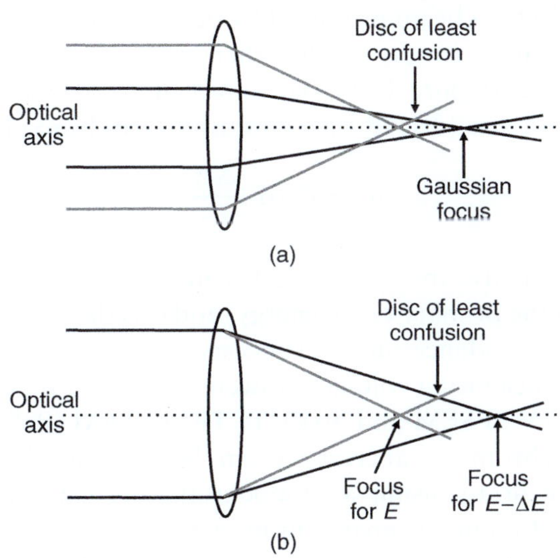

Figure 1.7 Schematic representation of (a) spherical and (b) chromatic aberration effects.

closer to the lens than the ones traveling at small angles. The desired focused spot thus becomes a disc stretched along the optic axis between the Gaussian focus and the lens, and called the disc of least confusion. The diameter of this disk (spot size) is (Cosslett, 1972):

$$d_s = 0.5 \cdot C_s \cdot \alpha^3 \tag{1.13}$$

where α is the convergence angle and Cs is spherical aberration coefficient. The minimum spot size (taking into account spherical aberration) therefore is proportional to α^3, while the current that can be delivered in the spot is proportional to α^2. These relationships force the user then to potentially sacrifice spot size for analytical performance or vice versa. In SEM spherical aberration can be minimized by working at very short working distances (in semi-in-lens systems Cs values can be brought down to 1–2 mm; for in-lens systems the Cs value can be as small as 50 μm - Joy, 2008) or by employing an aberration corrector (see Chapter 2). In TEM/STEM the spherical aberration effects are typically corrected via incorporation of higher order aberration corrector (Haider, 1995; Rose, 2009).

1.5.2 *Effect of Chromatic Aberration*

Chromatic aberration comes into play as a consequence of an energy spread in the electron beam produced by the electron source, resulting in longer focal distances for rays with higher energies and shorter distances for lower energies (Figure 1.7b). The effect of chromatic aberration on the resulting spot size (disc of least confusion) can be expressed as:

$$d_c = C_c \cdot \alpha \cdot \frac{\Delta E}{E_0} \tag{1.14}$$

where Cc is chromatic aberration coefficient, $\Delta E/E_0$ is the energy spread with respect to the primary beam energy, and α is the convergence angle. The spot size dependence on *1/E* suggests stronger effect of chromatic aberration as accelerating voltage is lowered, and becomes the dominant effect below 2 kV in SEM and 40 kV in TEM/STEM (Bell *et al.*, 2012). The effect of chromatic aberration can be minimized by employing a Cc corrector or by using a monochromator. Chapter 3 discusses incorporation of monochromator in modern SEM; Chapter 6 discusses the TEM monochromator.

1.5.3 *The Diffraction Limit*

The Airy disc is formed by the interference between the electrons traveling along the optical axis and the electrons scattered by the lens. The width of the beam d_d is a distance between the first-order zeroes in the Airy disc and is given by:

$$d_d = 0.6\,\frac{\lambda}{a} \tag{1.15}$$

and the deBroglie wavelength λ (in nm) for low energy electrons can be defined as:

$$\lambda = \frac{1.226}{\sqrt{E_0}} \tag{1.16}$$

where E_0 is the beam energy in eV. As the electron energy is lowered, the wavelength is increased, which then requires increase in the convergence angle to maintain negligible effect of diffraction effects on ultimate probe size. The schematic representation of the diffraction error is shown in Figure 1.8. Of course, other lens aberrations, such as coma, can be avoided by performing a coma free or on-axis alignment or by using small beam apertures and/or aberration corrected electron optics.

1.5.4 *Optimizing Spot Size for SEM and STEM*

The effects of all the aberration on the ultimate resolution that can be achieved in the electron microscope can be summed up as follows:

$$d_{eff}^2 = \left(0.6\frac{\lambda}{\alpha}\right)^2 + (0.5\cdot C_s\alpha^3)^2 + \left(C_c\cdot\alpha\frac{\Delta E}{E_0}\right)^2 + d_g^2 \tag{1.17}$$

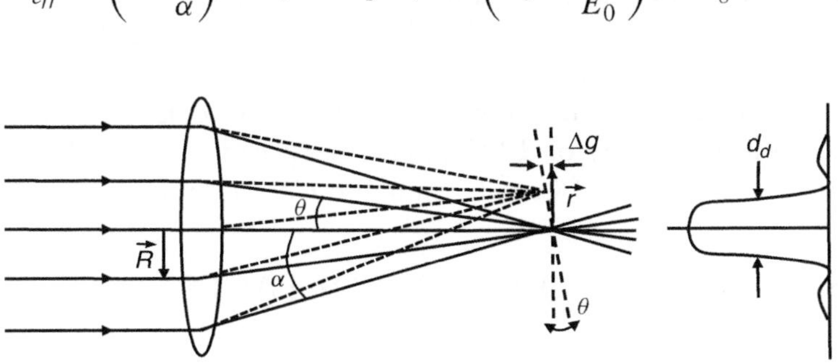

Figure 1.8 Schematic representation of the diffraction error.

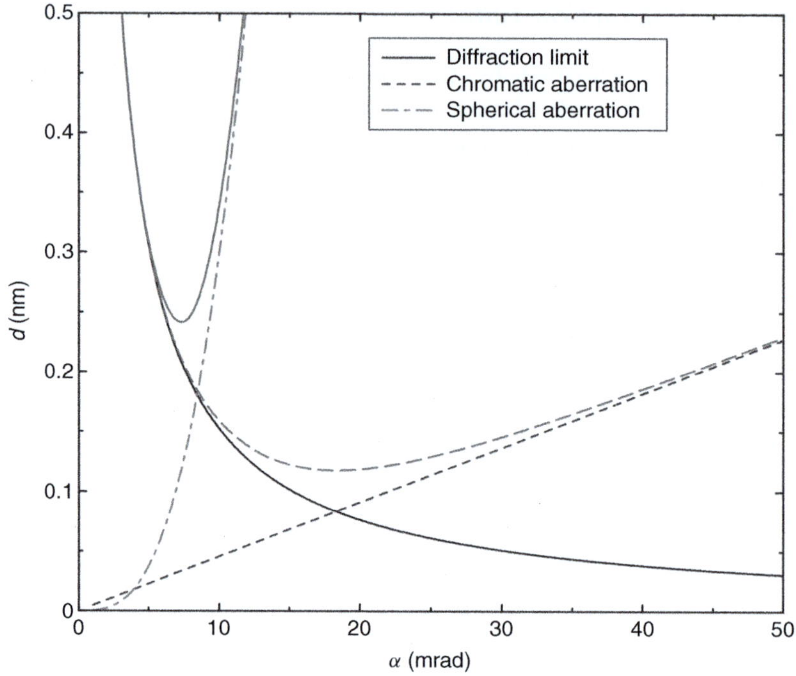

Figure 1.9 Effect of instrument aberrations on the probe size in electron microscope as a function of convergence angle α. The plot is for 200 kV, $\Delta E = 0.7$ eV, Cs = 1.2 mm, Cc = 1.3 mm.

The equation depicts a complex relationship between the effective probe size d_{eff}, the convergence angle and the accelerating voltage. The equation holds true for both scanning and transmission electron optical systems.

Figure 1.9 shows as an example a typical plot of spot size as a function of convergence angle for the Zeiss Libra 200 MC TEM. The plot indicates the limiting effects of instrument aberrations on the probe size in electron microscope as a function of convergence angle α for accelerating voltage of 200 kV, energy spread of $\Delta E = 0.7$ eV, and aberration coefficients of $Cs = 1.2$ mm and $Cc = 1.3$ mm .

1.6 SEM IMAGING AT LOW VOLTAGES

When scanning electron beam impacts a specimen in SEM, several different signals can be generated according to Figure 1.1. The individual signals are associated with a specific volume from which they are produced, and the range of the signals and their spatial distribution depend on the accelerating voltage as well as the type of material. Figure 1.10 shows a schematic diagram of the interaction volume of the primary electron beam with a specimen in SEM as well as the signal range. The energy spectrum in Figure 1.11 shows a representation of distribution of backscatter and secondary signals collected in the SEM. The spectrum shows a peak of elastically (low-loss) reflected BSEs, plasmon loss peaks

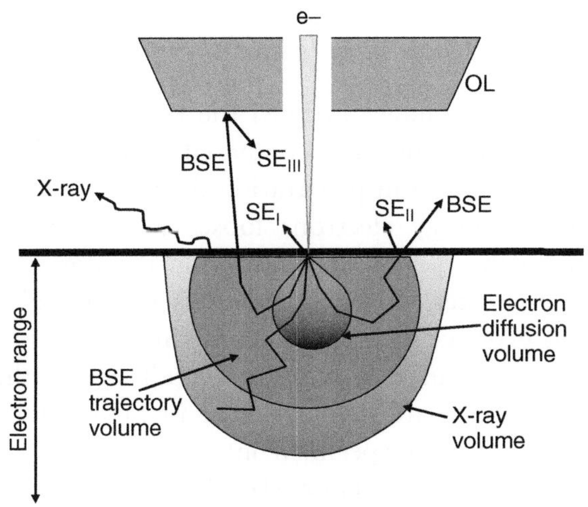

Figure 1.10 Beam/specimen interaction in SEM showing the resulting electron trajectories, generated signals and respective interaction volumes.

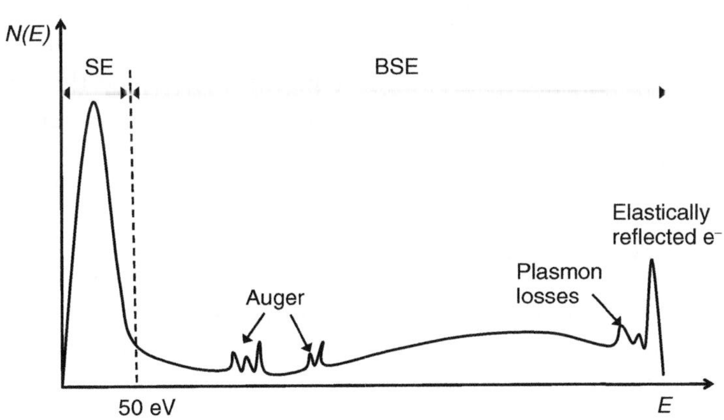

Figure 1.11 Energy spectrum N(E) (arbitrary units) of electron emission, showing secondary (SE), backscatter (BSE), Auger electrons regions.

and inner-shell ionization edges (inelastic scattering contribution), as well as Auger electrons (100–2000 eV) that can be generated from very thin surface layers. Next, we will discuss the basic concepts in SEM signal generation and detection, especially as they pertain to low voltage imaging. For further reading, there are various texts on SEM imaging and beam/solid interactions (e.g., Dekker, 1958; Seiler, 1983; Reimer, 1993; Joy and Joy, 1996; Joy, 1995; Goldstein *et al.*, 2003)

1.6.1 *Primary Contrast Signals and their Detection in SEM*

At lower accelerating voltages in SEM the beam/specimen interaction volume is substantially reduced as compared to high voltage operation, since the mean free path and penetration depth of the beam electrons are smaller (lower energy electrons loose energy faster). Moreover, the interaction volume is also dependent on the atomic number: the mean free path is longer in low Z materials versus high Z materials. High Z materials produce stronger Coulomb interactions (higher degree of elastic scattering), thus reducing the overall size of the interaction volume. These concepts are demonstrated in Figure 1.12 for carbon and gold, comparing beam/specimen interactions at 0.1 kV (assumed spot size is 5 nm) and 5 kV (spot size 2.5 nm). The image at 5 kV shows the familiar contrast where gold grains are white and the carbon support is black; however, at 0.1 kV the contrast is reversed and the gold grains have the same contrast as carbon. This phenomenon is easily explained by comparing beam specimen interactions at 0.1 kV versus 5 kV; at ultra-low voltages the interaction volumes are nearly identical. It becomes therefore apparent, that a priori knowledge of the material being analyzed in SEM is of outmost importance, since the choice of accelerating voltage will govern the understanding of the signals being produced and detected.

1.6.2 *Backscattered Electrons*

Backscattered electron signal is generated by the primary beam electrons that have undergone multiple elastic and inelastic collisions with specimen atoms and are reflected back through the specimen surface. The backscattering coefficient η is defined as the fraction of the primary electrons leaving the specimen with energies above 50 eV. The backscattering coefficient increases monotonically with incident electron angle and increasing atomic number Z, for accelerating energies

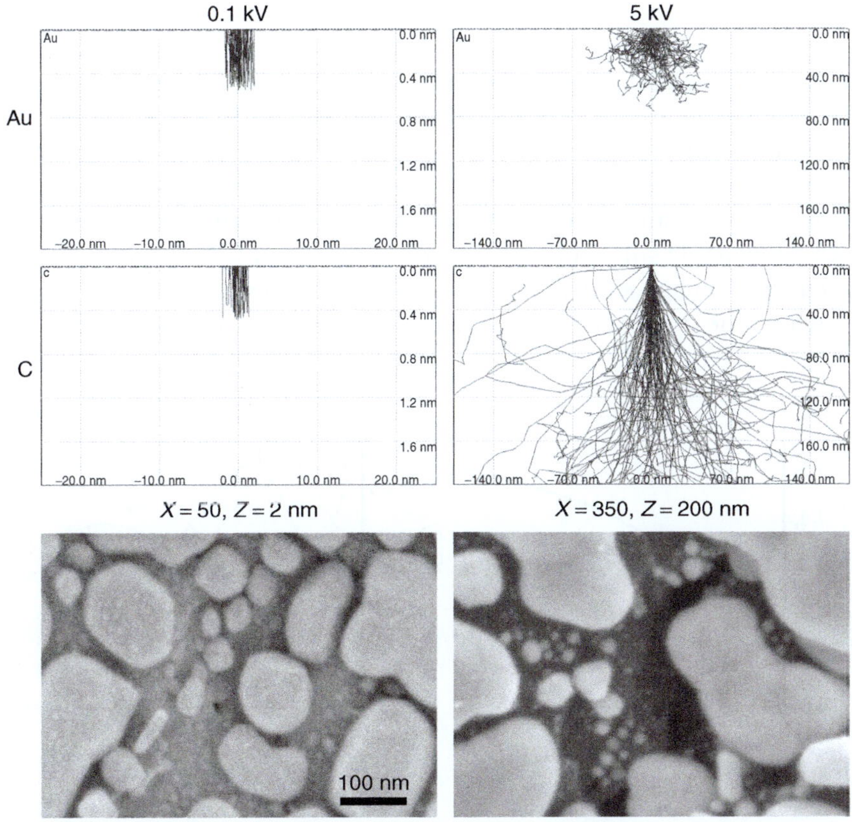

Figure 1.12 Monte Carlo simulation (CASINO®) of electron trajectories in Au and C as a function of accelerating voltage (top). Corresponding images are shown at the bottom.

above 5 kV, Figure 1.13(a) (Darlington, 1975; Drescher, *et al.*, 1970; Goldstein *et al.*, 2003):

$$\eta(Z,\phi) = 0.89\cdot\left(\frac{\eta(Z,0)}{0.89}\right)^{\cos\phi} \tag{1.18}$$

$$\eta(Z,0) = -0.0254 + 0.016.Z - 1.86.10^{-4}\cdot Z^2 + 8.3\cdot 10^{-7}\cdot Z \tag{1.19}$$

where ϕ is the specimen tilt angle. However, the BSE coefficient behavior changes substantially at low voltages, mainly due to the Mott elastic cross-section, as demonstrated in Figure 1.13(b). As a general trend, the plot shows that η increases with decrease in energy for $Z<30$ and decreases for $Z>30$ at low accelerating voltages (less than 5 kV)

(Darlington and Cosslett, 1972; Reimer and Tollkamp, 1980; Kotera *et al.*, 1981a; Kotera *et al.*, 1981b; Schmid *et al.*, 1983). Importantly, the trend of the plot suggests that at low voltages the BSE image contrast can be somewhat complicated and the user should be extremely careful about image interpretation.

An example of this complicated contrast phenomenon is shown in Figure 1.14; a sample of Ni/Pt eutectic with AlN spheres was imaged with a low angle solid-state backscatter detector at 0.5, 2 and 5 kV. Although 5kV image shows distinct contrast differences between Ni and Pt while the lower Z AlN spheres have darker contrast, reducing

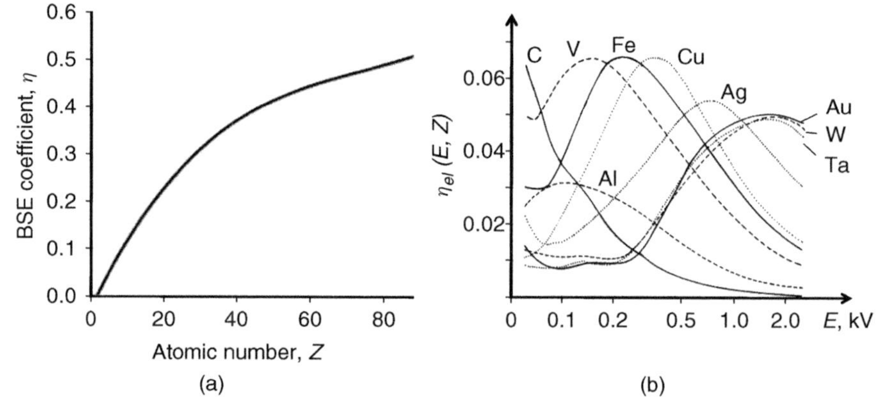

Figure 1.13 (a) Backscatter coefficient as a function of atomic number, Z; (b) Elastic backscatter coefficient for different materials as a function of accelerating voltage (after Schmid *et al.*, 1983).

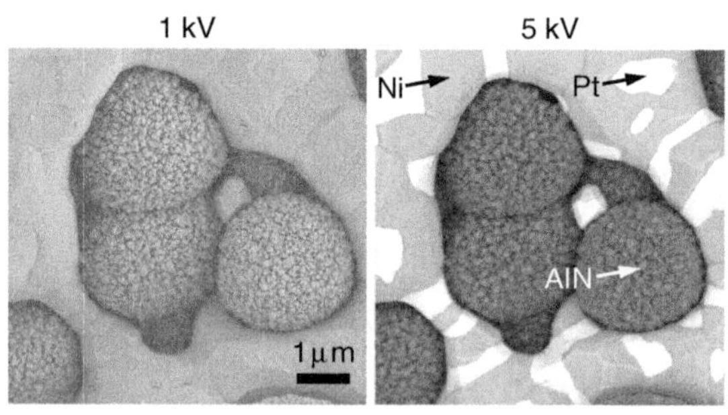

Figure 1.14 An example of BSE contrast reversal upon lowering accelerating voltage. Sample: ceramic eutectic. The images were taken with low angle solid state BSE detector (JEOL USA).

the accelerating voltage down to 0.5 kV essentially eliminates contrast differences between Ni and Pt as well as metals and ceramic components.

1.6.3 *Secondary Electrons*

The SE yield (δ) is defined as the mean number of secondary electrons excited per incident electron. The probability of the secondary electrons escaping from a surface can be given by:

$$p(z) = 0.5\exp\left(\frac{-z}{\lambda}\right) \tag{1.20}$$

where λ is a mean free path (or escape depth) of the electrons propagating to the surface. Seiler (1983) demonstrated that λ is approximately 1 nm in metals and up to 20 nm in insulators. The maximum depth of emission is approximately 5λ. Traditionally, the electrons emitted with energies below 50 eV are considered to be secondary electrons (see Figure 1.11), and their energy spectrum distribution is independent of accelerating voltage (if E_{acc}>0.1 kV). The secondary emission can be characterized by a curve with a peak at approximately 2–5 eV (Figure 1.15a); the peak's general shape is essentially the same for various materials. However, the width of the peak as well as its position vary between different materials and with presence of surface contamination layers; the width of the peak is narrower and its position shifts towards lower energies for insulators as compared to metals (Bouchard and Carette, 1980; Seiler, 1983; Cazaux, 2008). A typical value for δ is ~0.1 for most elements; the value for Au is 0.2 and for C is 0.05. The maximum SE yield can

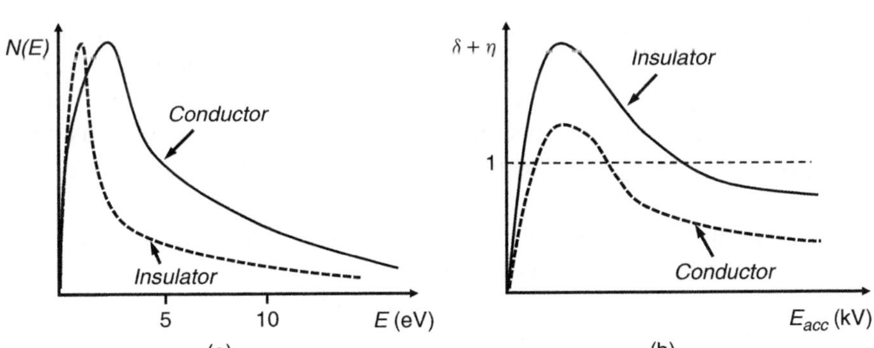

Figure 1.15 (a) Typical energy distribution of SE electrons from conductors and insulators; (b) Total electron yield as a function of primary electron energy.

be achieved when the electron range in the sample is approximately the escape depth of the secondary electrons.

Secondary electrons are further divided into three categories, based on their origin. SE_I electrons are generated within the top 1–10 nm of the specimen, and are the only electrons that contribute to high resolution on the order of electron probe size. SE_{II} electrons are produced by backscattered electrons within the interaction volume on the order of 0.1–1 μm in diameter at high kV, and 5–50 nm at low kV (depending on the material). For high Z materials, SE_{II} signal is approximately 1.5 times larger than SE_I signal; for low Z materials the ratio of SE_{II} to SE_I is approximately 1/5. SE_{III} signal is what is often referred to as "system electrons", originating from BSEs interacting with the parts of the SEM chamber or most likely with the bottom of the objective lens and producing SE signal. In certain semi in-lens systems, conversion to SE_{III} electrons is used to boost the amount of SEs that can be detected by the in-lens detector, especially at low voltages.

Taking into account both secondary and backscattered electron emission, the total electron yield is therefore:

$$\delta = \delta_{SE} + \delta_{BSE} \tag{1.21}$$

Comparison of the total electron yield behavior for metals and insulators is shown in Figure 1.15(b). At high kVs (above 5 kV), δ is below unity; however, as the accelerating voltage is lowered, the total yield increases until it reaches a crossover point E_2 where $\delta = 1$. Further reduction in accelerating voltage leads to a maximum in total yield ($\delta > 1$), and follows by a decrease to another crossover point E_1 where the total yield $\delta = 1$ again. This behavior can be explained by significant increase in SE yield between the two crossover points due to the decrease in interaction volume as a function of accelerating voltage; the shallow interaction volume between E_1 and E_2 results in larger number of electrons within the escape depth of the specimen able to leave the surface as SEs. This phenomenon should be taken into consideration when selecting the imaging conditions for heterogeneous materials; at low accelerating voltages the image may show reversal of contrast based on the choice of voltage with respect to points E_1 and E_2.

The SE yield (similar to backscattered yield) is also dependent on the specimen tilt with respect to the primary beam, and follows a secant law, $\delta(\theta) = \delta_0 \sec(\theta)$. This dependence is demonstrated in Figure 1.16. Similar considerations also come into play when considering sample features like corners, steps or edges, which inevitably increase SE emission.

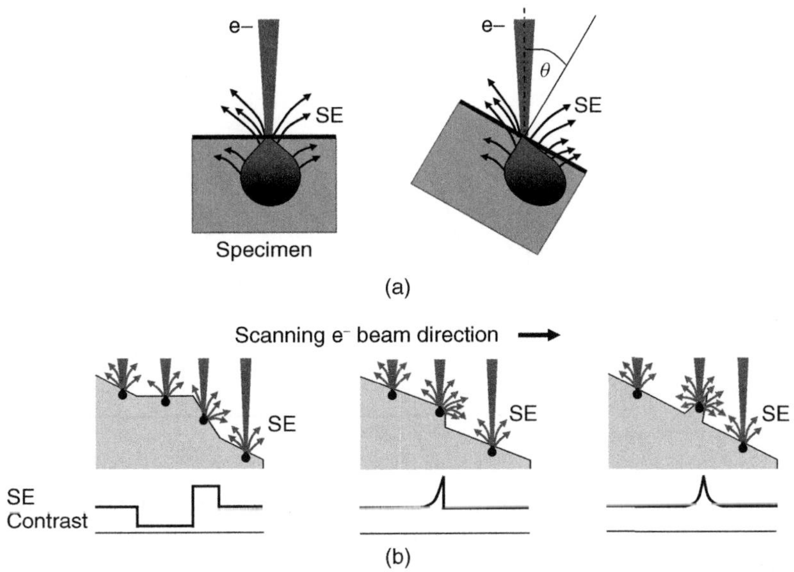

Figure 1.16 (a) Secondary electron yield as a function of tilt angle with respect to the primary beam. (b) Schematic representation of the resulting detected SE contrast (E/T detector) as a function of specimen topography.

1.6.4 *Charge Balance in SEM*

Interaction of primary beam with insulating specimen can produce a surplus or deficit of electrons on the specimen surface that results in the buildup of unstable and inhomogeneous electric fields and leads to charging and inability to image the specimen. If the accelerating voltage is chosen to be equal to E_1 or E_2 crossover energies, the number of incident electrons equals the number of emitted electrons (SE and BSE), thus the net charge is zero. The regions where $E<E_2$ or $E>E_1$ are considered regions of "white" charging; when the accelerating voltage is chosen to be $E_1<E<E_2$ the specimen exhibits "black" charging. Up till recently it was fairly straightforward to image in the region between E_1 and E_2; however, going below E_1 for most materials has been very difficult due to instrument limitations. The new instrument designs that allow imaging down to few tens of volts with fairly high resolution can alleviate these concerns. Figure 1.17 shows an example of pollen grain imaged in the "black" contrast ($E_1<E_{acc}<E_2$) region and ultra-low voltage ($E<E_1$) region with contrast reversal and minimal charging. Additional information regarding crossover energies values for various

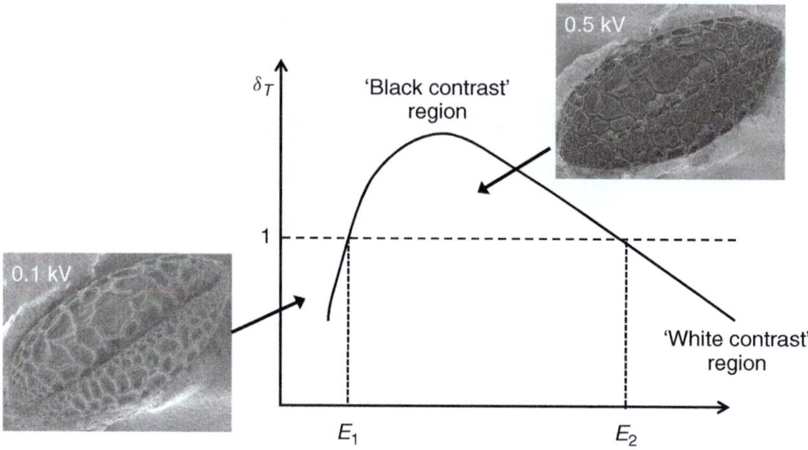

Figure 1.17 Total electron yield as a function of energy, showing with regions of possible contrast reversal. Images of pollen grain at various corresponding voltages shown as insets.

materials can be found in Seiler (1983), Joy and Joy (1996), Goldstein *et al.* (2003).

Low voltage imaging can be particularly useful for charge reduction. However, care should be taken in the case of imaging of rough specimens as well as composites of insulating and conducting materials, since charge dissipation may not be as straightforward because of the complex nature of the electric fields generated on the specimen surface. In addition to lowering the accelerating voltage for charge reduction, the user may consider lowering the probe current as well as changing scan speed to acquire the image (in particular, use line or frame integration). Chapter 2 will discuss the strategies employed in newly designed FE-SEM instruments for charge compensation and imaging of insulating materials.

1.6.5 *SEM Image Contrast*

One of the main advantages of low accelerating voltage SEM is the ability to tap into different types of contrast mechanisms that are only available through changes in electron range, secondary electron yield and backscatter electron coefficients of elements due to low kV operation. The main types of contrast mechanisms that undergo substantial changes as compared to the more traditional high kV observation are listed next:

- Topography
- Material (compositional) contrast
- Crystal orientation or channeling

- Thin film, coating or contamination layer contrast
- Voltage contrast
- Magnetic domain contrast

As was shown previously by Monte-Carlo simulations, there is a substantial difference in interaction volume between SE and BSE signals at higher accelerating voltages; at low kVs these differences are dramatically reduced, which leads to some very interesting reversals of traditionally expected topographic and material contrast as well as added benefits for surface imaging. Further discussion and examples of the contrast mechanisms at low voltages can be found in (Cazaux, 2004; Cazaux, 2008). In Chapters 2 through 4 we will discuss examples of a variety of material and biological specimens imaged and analyzed at low voltages, showing distinct advantages of low kV operation.

1.6.6 *Microanalysis in SEM at Low Voltages*

One of the major benefits of low voltage microanalysis is increased spatial resolution through the reduction of beam/specimen interaction volume; moreover, lower accelerating voltage may allow longer analysis times for beam sensitive specimens. An example of thin section of ore imaged and analyzed at various accelerating voltages is shown in Figure 1.18; there is an obvious benefit to operating at 5 kV versus 15 kV to resolve the fine material composition and microstructure. Microanalysis at lower voltages can determine chemical composition of thin films, nanoparticles, surface layers and light elements (Boyes, 2000; Rowlands *et al.*, 2009).

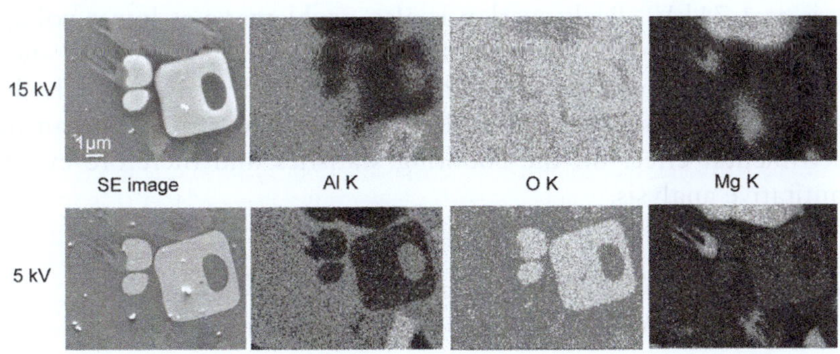

Figure 1.18 Advantage of X-ray microanalysis at low voltages in SEM. WDS maps of polished section of ore from Miake Island. Elemental maps of Al, O and Mg shown alongside corresponding SE images, taken at 15 kV (top) and 5 kV (bottom) (See plate section for coloured version).

As with SEM imaging, the microanalysis techniques (such as EDS, WDS and CL) have focused in the past on the use of high accelerating voltage, mainly due to the requirement for adequate X-ray signal detection without the need to spend hours and even days performing a single analysis. The ability to collect X-ray signal with all the characteristic lines present, thus (hopefully) eliminating all the potential issues associated with element lines overlaps has also driven researchers towards higher accelerating voltages. However, the most important and limiting factor for low voltage microanalysis until recently has been the fact that a relatively large probe size at low accelerating voltage and high beam current necessary for analysis prevented the user from taking advantage of the reduced interaction volume and thus better spatial resolution provided by low kV operation. This paradigm has changed in the recent years with the development of FEG-SEMs that boast nanometer size resolution at 1 kV, and maintain relatively small probe size even when beam current is increased into several nA range necessary for microanalysis (see Chapter 2). The new advances with the EDS detector design and analysis software improvements (e.g., Kenik *et al.*, 2004; Collins *et al.*, 2009; Newbury, 2008) have also aided in improvements in X-ray analysis at low voltages. Additional benefit of lower kV is the reduction in X-ray absorption; this suggests that the quantitative analysis at lower voltages should be more accurate than at high kV for the same number of X-ray counts in the spectrum (Gauvin, 2006).

The low voltage X-ray analysis, however, comes with some caveats. Some elements are difficult to detect at low voltages, and the user should be careful in the selection of the accelerating voltage. For instance, if the material of interest is silicon, the user requires at least 3.5 kV accelerating voltage to get significant counts for X-ray analysis (Si K line is at 1.74 kV). Peak overlap is also problematic at lower voltages, where multiple elements have K, L and M X-ray lines. The user should also be careful about specimen contamination and hydrocarbon buildup on the specimen surface during low voltage analysis, which can have a detrimental effect on the counting statistics and therefore eventual quantitative analysis.

1.7 TEM/STEM IMAGING AND ANALYSIS AT LOW VOLTAGES

The main driving force behind resurgence of low voltage transmission electron microscopy in the recent years is the potential to significantly reduce the knock-on beam damage and thus improve the ability to effectively observe and characterize at atomic scale new types of nanomaterials. Another important advantage of low voltage TEM and STEM is enhanced contrast due to energy dependent scattering behavior.

The incorporation of aberration correctors, in particular the successful adoption of C_s correctors in TEM columns has allowed the increased resolution now possible at sub-Ångström levels. In the process of improving the resolution at high accelerating voltages, that is, above 200 kV, the C_s corrector has also enabled the improvement of resolution at lower energies (<120 keV). Low-voltage High-Resolution Electron Microscopy (LVHREM) has several advantages: increased cross-sections for inelastic and elastic scattering and hence higher contrast efficiency from each atom, as well as reduced radiation knock-on damage to samples insensitive to other damage mechanisms, which includes most metals, semiconductors and other solid state materials.

Historically, the use of higher TEM voltages was favored since they reduce the spherical and chromatic aberration effects; however, the development of spherical aberration correctors has allowed atomic resolution at 60 kV (Krivanek *et al.*, 2008).

Although the TEM samples must be significantly thinner for low kV observation, the improvement in contrast for inorganic materials, organics, biological samples and especially nanobiological samples in low-voltage TEM while retaining atomic resolution cannot be understated. Damage mechanisms for biological samples are complicated and very structure dependent.

Lower beam energy is also important for obtaining the highest possible energy stability for spectroscopy applications. At 40 kV using an electron monochromator a beam with energy distribution at a full width at half maximum of less than 50 meV can be obtained. By reducing the voltage of the TEM, there are distinct gains in using electron energy-loss spectroscopy to determine band gaps and the dielectric properties with nanometer spatial resolution: the relativistic losses are reduced and delocalization of the energy-loss signal is reduced with the energy of the incident probe. Since C_c is both expensive and difficult to correct, the resolution improvement by using a monochromator is an effective option for materials applications. Details of low voltage TEM will be described in a Chapter 5 in this volume. Applications of low voltage STEM and EDS will be presented in Chapters 6 and 7.

1.8 CONCLUSION

There are significant advantages to low voltage operation across the different electron optical platforms. In the case of SEM, the significant reduction in interaction volume at lower kVs allows the user to obtain surface sensitive information that would be impossible to observe at high kV. Low voltage operation also promotes imaging on non-conductive specimens without the need for conductive coating, and imaging of beam sensitive materials without inducing extensive radiation damage. Furthermore, changes in the secondary and backscatter yields open avenues for new and previously untapped contrast mechanisms. There are very distinct differences between the important considerations for low voltage SEM imaging as compared to TEM and STEM imaging and these will be detailed in the later chapters of this volume.

REFERENCES

D.C. Bell, C.J. Russo, and D.V. Kolmykov, 40 keV atomic resolution TEM, *Ultramicroscopy*, **114**, 31–37 (2012).

G.A. Botton, S. Lazar, and C. Dwyer, Elemental mapping at the atomic scale using low accelerating voltages, *Ultramicroscopy*, **110**, 926–934 (2010).

E.D. Boyes, On low voltage scanning electron microscopy and chemical microanalysis, *Microsc. Microanal.*, **6**, 307–316 (2000).

C. Bouchard and J.D. Carette, The surface potential barrier in secondary emission from semiconductors, *Surf. Sci.*, **100**, 241–250 (1980).

R. Browning, T.Z. Li, B. Chui, *et al.*, *Low-energy electron/atom elastic scattering cross sections* from 0.1–30 keV, *Scanning*, **17**, 250 (1995).

J. Cazaux, Charging in scanning electron microscopy "from inside and outside", *Scanning*, **26**, 181–203 (2004).

J. Cazaux, On some contrast reversals in SEM: application to metal/insulator systems, *Ultramicroscopy*, **108**, 1645–1652 (2008).

C. Colliex and C. Mory, Quantitative aspects of scanning transmission electron miscoscopy, in *Quantitative Electron Microscopy*, J.N. Chapman and A.J. Craven (eds.), Scottish Universities Summer School in Physics, Edinburgh, 149–216 (1984).

C.L. Collins, J. Holland, S.R. Burgess, and N. Rowlands, X-Max Large Area SDD Detectors - Creating a Real Impact on Nano-Science, *Microsc. Microanal.*, **15**, 172–173 (2009).

V.E. Cosslett, Probe size and probe current in the STEM, *Optic*, **36**, 85–92 (1972).

A.J. Dekker, Secondary electron emission, in *Solid State Physics*, F. Seitz and D. Turnbull (eds.), Academic Press, Inc., 251–311 (1958).

Z. Czyzewski, D.O. MacCallum, A. Romig, and D.C. Joy, Calculations of Mott scattering cross-section, *J. Appl. Phys.*, **68**, 3066 (1990).

E.H. Darlington, Backscattering of 10–100 keV electrons from thick targets, *J. Phys. D*, **8**, 85–93 (1975).

E.H. Darlington and V.E. Cosslett, Backscattering of 0.5–10 keV electrons from solid targets, *J. Phys. D*, **5**, 1969–1981 (1972).

H. Drescher, L. Reimer, and H. Seidel, Backscattering and secondary electron emission of 10–100 keV electrons and correlations to scanning electron microscopy, *Z. Ang. Physic*, **29**, 331–338 (1970).

D. Drouin, A.R. Couture, D. Joly, *et al.*, CASINO V2.42—A fast and easy-to-use modeling tool for scanning electron microscopy and microanalysis users, *Scanning*, **29**, 92–101 (2007).

R.F. Egerton, *Electron Energy Loss Spectroscopy in the Electron Microscope*, 3rd edn, Plenum Press, New York, (2011).

R. Gauvin and D. Drouin, A formula to compute total elastic Mott cross-sections, *Scanning*, **15** (3), 140–150 (1993).

R. Gauvin, K. Robertson, P. Horny, *et al.*, Materials characterization using high-resolution scanning electron microscopy and X-ray microanalysis, *JOM*, **58** (3), 20–25 (2006).

J. Goldstein, D.E. Newbury, D.C. Joy, *et al.*, *Scanning Electron Microscopy and X-Ray Microanalysis*, 3rd edn, Springer, New York, (2003).

M. Haider, G. Braunshausen, and E. Schwan, Correction of the spherical-aberration of a 200-kV TEM by means of a hexapole-corrector, *Optik*, **99**, 167 (1995).

P.W. Hawkes and E. Kasper, *Principles of Electron Optics*, vol. **1**, Academic Press, London (1985).

L.W. Hobbs, *Introduction to Analytical Electron Microscopy*, p. 437, J.J. Hren, J.I. Goldstein, and D.C. Joy (eds), Scanning Microscopy International, Chicago (1979).

L.W. Hobbs and M.R. Pascucci, Radiolysis and defect structure in electron-irradiated α-quartz, *J. Phys. Colloques*, **41**, 237 (1980).

P. Hovington, D. Drouin, R. Gauvin, *et al.*, CASINO: A new Monte Carlo code in C language for electron beam interactions—Part III: Stopping power at low energies, *Scanning*, **19**, 29–35 (1997).

D. Joy, A database on electron-solid interactions, *Scanning*, **17**, 270–275 (1995).

D. Joy, Aberration-Corrected SEM, in *Biological Low-Voltage Scanning Electron Microscopy*, H. Schatten and J.B. Pawley (eds), Springer, Heidelberg (2008).

D.C. Joy and C.S. Joy, Low Voltage Scanning Electron Microscopy, *Micron*, **27**, 247–263 (1996).

E.A. Kenik, D.C. Joy and D. Redfern, Microcalorimeter detectors and low voltage SEM microanalysis, *Microchim. Acta*, **145**, 81–85 (2004).

M. Knoll, Aufladepotentiel und Sekundäremission elektronenbestrahlter Körper, *Z. tech. Phys.*, **16**, 467–475 (1935).

M. Kotera, K. Murata, and K. Nagami, Monte Carlo simulation of 1–10 keV electron scattering in a gold target, *J. Appl. Phys.*, **52**, 997–1003 (1981a).

M. Kotera, K. Murata, and K. Nagami, Monte Carlo simulation of 1–10 keV electron scattering in an aluminum target, *J. Appl. Phys.*, **52**, 7403–7408 (1981b).

O.L. Krivanek, N. Dellby, R.J. Keyse, *et al.*, Aberration-corrected STEM and EELS, in *Advances in Imaging and Electron Physics*, P.W. Hawkes (ed.), Elsevier, Amsterdam (2008).

F. Lenz, Zur Streuung mittelschneller Elektronen in kleinste Winkel, *Z. Naturforsch.*, **9A**, 185–204 (1954).

H. Mahl, Supermicroscopic determination of the orientation of single aluminium crystals (in German), *Metallwirtschaft*, **19**, 1082–1085 (1940).

D. McMullan, *Investigations relating to the design of electron microscopes*, Ph.D. dissertation, University of Cambridge, (1952).

D. McMullan, A history of the scanning electron microscope, 1928–1965, *Adv. Imaging Electron Phys.*, **133**, 523–545 (2004).

N.F. Mott, The polarisation of electrons by double scattering, *Proc. R. Soc. London A*, **135**, 429 (1932).

N.F. Mott and H.S.W. Massey, *The Theory of Atomic Collisions*, Clarendon Press, Oxford, (1949).

D.A. Muller, L. Fitting-Kourkoutis, M. Murfitt, *et al.*, Atomic-scale chemical imaging of composition and bonding by aberration-corrected microscopy, *Science*, **319**, 1073 (2008).

D.E. Newbury, Developments in instrumentation for microanalysis in low-voltage scanning electron microscopy, in *Biological Low-Voltage Scanning Electron Microscopy*, H. Schatten and J.B. Pawley (eds), Springer, Heidelberg (2008).

M.R. Pascucci, J.L. Hutchison, and L.W. Hobbs, The metamict transformation in alpha-quartz, *Radiat. Eff.*, **74**, 219 (1983).

L. Reimer, *Image Formation in Low-Voltage Scanning Electron Microscopy, Vol. TT 12*, SPIE, Bellingham, WA (1993).

L. Reimer and H. Kohl, *Transmission Electron Microscopy, Physics of Image Formation and Microanalysis*, Springer, Berlin (2008).

L. Reimer and C. Tollkamp, Measuring the backscattering coefficient and secondary electron yield inside a SEM, *Scanning*, **3**, 35–39 (1980).

H. Rose, *Geometrical Charged Particle Optics*, Springer, Berlin/Heidelberg (2009).

N. Rowlands, N. Erdman and V. Robertson, Characterization of nanoscale structures using a combination of an ultra-high resolution field emission SEM and large area silicon drift detector, *Microsc. Microanal.*, **15**, 548–549 (2009).

E. Ruska, The electron microscopic imaging of surfaces of surfaces irradiated with electrons (in German), *Z. Phys.*, **83**, 492–497 (1933).

E. Rutherford, Elastic cross section, *Phil. Mag.*, **21**, 669–680 (1911).

R.F.M. Thornley, *New applications of the electron microscope*, Ph.D. dissertation, University of Cambridge, (1960).

R. Schmid, K.H. Gaukler, and H. Seiler, Measurement of elastically reflected electrons ($E \leq 2.5$ keV) for imaging of surfaces in a simple ultrahigh vacuum SEM, *Scanning Electron Microscopy 1983/II*, SEM Inc., AMF O'Hare, 501–509 (1983).

H. Seiler, Secondary electron emission in scanning electron microscope, *J. Appl. Phys.*, **54**, R1–R18 (1983).

O. Ugurlu, J. Haus, A.A. Gunawan, *et al.*, Radiolysis to knock-on damage transition in zeolites under electron beam irradiation, *Phys. Rev. B*, **83**, 113408 (2011).

M. von Ardenne, *Improvements in electron microscopes*, British Patent No. 511204 (1937).

M. von Ardenne, The scanning electron microscope. Theoretical fundamentals (in German), *Z. Phys.*, **109**, 553–572 (1938).

M. von Ardenne, On the history of scanning electron microscopy, the electron microprobe, and early contributions to transmission electron microscopy, in "The beginnings of electron microscopy" (P.W. Hawkes, ed.), *Adv. Electonics Electr. Phys.*, **Suppl. 16**, 1–21 (1985).

O.C. Wells, *The construction of a scanning electron microscope and its application to the study of fibres*. PhD Dissertation, Cambridge University (1957).

V.A. Zworykin, J. Hillier, and R.L. Snyder, A scanning electron microscope, *ASTM Bull.*, **117**, 15–23 (1942).

2

SEM Instrumentation Developments for Low kV Imaging and Microanalysis

Natasha Erdman[1] and David C. Bell[2]
[1]*JEOL USA Inc., USA*
[2]*School of Engineering and Applied Sciences, Harvard University, USA*

2.1 INTRODUCTION

Scanning electron microscopes have historically been operated at relatively high voltages, between 15–40 kV; however, recent instrumental developments have now enabled very much the possibility for ultra-low voltage imaging in the order of tens of volts of accelerating voltage. The benefits of low voltage microscopy in SEM for observation of non-conductive specimens and high spatial resolution combined with enhanced surface detail have been known since late 1960s (Thornley, 1960; Thornley and Cartz, 1962), the capability for good SEM resolution at lower voltages (sub 5 kV) was extremely lacking at that time. The rapid growth of the semiconductor industry in the early 1980s dictated the development of new SEM columns with high-resolution across a range of accelerating voltages (in particular the 1 kV range) that could accommodate handling of large wafer semiconductor specimens.

The fundamental limitations to the operation of the scanning electron microscope, in particular in the low voltage regime, have been outlined early on in the development of SEMs (Oatley *et al.*, 1965):

Low Voltage Electron Microscopy: Principles and Applications, First Edition.
Edited by David C. Bell and Natasha Erdman.

1. *Brightness of the electron source.* According to Langmuir (1937), the maximum current density for electrons generated by thermionic cathode emission and brought to a focus by an electron-optical system can be described by the following equation:

$$J = J_0 \cdot \left(\frac{E}{kT} + 1 \right) \cdot \sin^2 \alpha \approx J_0 \cdot \frac{E \cdot \alpha^2}{kT} \tag{2.1}$$

   where J_0 is the current density at the surface of the thermionic cathode, E is the accelerating voltage applied to the source, k is the Boltzman constant, T is temperature of the electron source and α is the convergence angle. Since E is much larger than kT and α is always small, the equation can be approximated accordingly. Thus, work in the low kV regime results in a significantly larger spot size (and hence reduced resolution) to maintain the same level of beam current and source brightness.
2. *Lens aberrations.* The nature of lens aberrations in the electron optical system and their effects on image resolution is discussed in Chapter 1. It is important, though, to point out that the effect of chromatic aberration on the disc of least confusion scales with $\alpha \Delta E/E$, meaning that at lower accelerating voltages it is imperative to maintain small convergence angle α in order to maintain best image resolution. This, of course, becomes a contradicting argument to the previous point about maintaining high brightness and probe current (better signal to noise ratio in the images). These seemingly negating needs are further compounded by the fact that decreasing the convergence angle also increases the influence of diffraction on the ultimate spot size. The eventual solution in practical terms is to balance all these effects for the best possible outcome at any particular time.
3. *Wavelength of the electrons.* Naturally from the DeBroglie relationship, reducing the accelerating voltage of the source increases the wavelength of the electrons, thus inherently giving us a larger probe size than at a higher kV. Increased wavelength enhances effects of the diffraction contribution to the size of the disc of least confusion ($d_d = 1.22\lambda/\alpha$); therefore, the convergence angle α should be increased at lower kVs to maintain the smallest spot size.
4. *Environmental influences.* Operation at lower kVs at high resolution requires improved stability of the instruments, overall instrument cleanliness (improved vacuum systems) and proper magnetic shielding. In particular, the influence of stray magnetic fields is proportional

to $E^{-1/2}$ (Mullerova and Fiser, 1994). In the recent generation of instruments, specimen cleanliness and lack of hydrocarbon (or other) contamination layers on the surface plays a crucial role in the ability to obtain best possible SEM images at ultra-low kV.

These considerations are still relevant today and have driven the vast improvements in the design and performance of SEMs over the last two decades. The instrumentation has improved to a point where in FE-SEM the nominal resolution at 30 kV is on the order of 0.8 nm (for bulk specimens) and 0.6 nm for thin membranes in semi-in-lens systems (0.4 nm for a bulk specimen positioned inside the objective lens); the resolution at 1 kV is 1.4 nm and 5 nm at 0.1 kV. This chapter highlights the current state of the art instrumentation as well as different approaches taken by the SEM manufacturers to improve low kV performance of the current systems.

2.2 THE ELECTRON SOURCE

Over the years four possible choices of electron sources for the SEM have emerged—thermionic W and lanthanum hexaboride (LaB_6) cathodes and point emission cathodes, ZrO/W Schottky (also referred to as 'thermal field emission') and cold field emission (see for example Vladar and Postek, 2008). Figure 2.1 shows a schematic of the different electron sources.

The traditional source used from the inception of the SEM instruments has been thermionic cathode W filament. Thermionic emitters have to work with relatively high cathode temperature ($T_c = 2500–3000$ K, T_m for W is 3650 K) in order to overcome the work function barrier and produce emission. These high temperatures limit the lifetime of W filaments to ~50–100 h on the average. The energy spread of the thermionic source is on the order of 1–2 eV. In the 1970s, LaB_6 thermionic cathodes became popular (Broers, 1974), in particular due to improved source brightness (10–20 times higher) and longer lifetime (about 10 times) than the traditional W filaments.

Schottky emitters are made of ZrO_2 coated W(100) with a tip radius of 0.5–1 μm. The presence of ZrO_2 coating lowers the work function of W from 4.6 eV to 2.6 eV, which subsequently lowers the temperature at which the source is operated. Even though the field strength required to extract electrons from a Schottky source is smaller than what is needed to achieve a tunneling effect, the consensus has been to refer to Schottky-type sources as 'field emission'. The source lifetime is on the

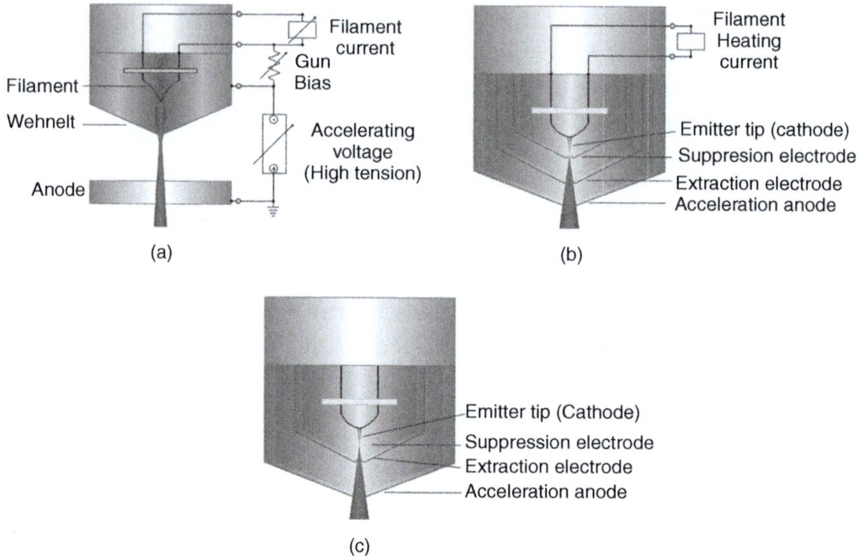

Figure 2.1 Comparison of different electron sources: (a) W filament; (b) Schottky FEG; (c) cold cathode FEG.

average 3–5 years (with good vacuum system); the emission stops when the reservoir of ZrO_2 is depleted. The current density of the Schottky source is 2–3 times larger than the thermionic source, and the energy spread is around 0.5–1.0 eV. These characteristics make the Schottky source the most popular choice for both high resolution and analytical work in SEM (Swanson and Tuggle, 1981; Tuggle and Swanson, 1985).

Crewe *et al.* (1968) introduced a working cold field emission cathode, which had much more stringent vacuum requirements, but provided 10^5 times higher brightness than the W filaments and had a significantly narrower energy spread, around 0.2–0.5 eV. In cold field emission gun the emission occurs via tunneling effect, by increasing the electric field at the cathode to more than $10^8\,V\,cm^{-1}$ so that the width of the potential barrier becomes less than 10 nm. The tips used for cold-cathode sources are typically either <100> or <310> oriented W; they operate at room temperature but need flashing (heating) daily to remove the contamination layer (gas molecules) that accumulates on the tip and subsequently reduces emission. The lifetime of cold-cathode tip is highly dependent on the cleanliness of the vacuum system, with some tips lasting over three years. The main advantages of cold-cathode instruments are high brightness and narrow energy spread (particularly important in low voltage SEM imaging, since it reduces the effects of chromatic aberration); the main drawbacks are the current stability and the relatively low probe current (standard instrument settings allow up to 20 nA depending on the accelerating voltage selection; 64 nA can be achieved, but significantly reduces the source stability).

The discussion in this chapter will focus on the design and operation of SEMs with field emission sources (Schottky and cold cathode types). The size of the virtual source, the resulting probe size at the specimen surface as well as the source brightness imply that FE-SEM is the best type of instrument for operating in the low kV regime with high image resolution.

In the last decade there have been several improvements in the field emission electron gun technology. One of the approaches used to increase the amount of probe current delivered to the specimen at any given kV is to use 'in-lens' electron gun (JEOL Ltd). The schematic of such a gun is shown in Figure 2.2. The electron source is placed inside the magnetic field of a low aberration condenser lens, which is integrated with the anode of the FEG; the field-emitted electrons are constricted into a parallel beam as they accelerate. The high brightness electron beam is confined into a narrow angle, and passes through the anode aperture, and then is converged into a small spot by the subsequent objective, thus creating a system capable of delivering 10 times more current (200–400 nA) than the conventional FEG systems. An addition of an aperture angle control lens (ACL) incorporated in the objective lens design ensures that not only can such gun deliver a high probe

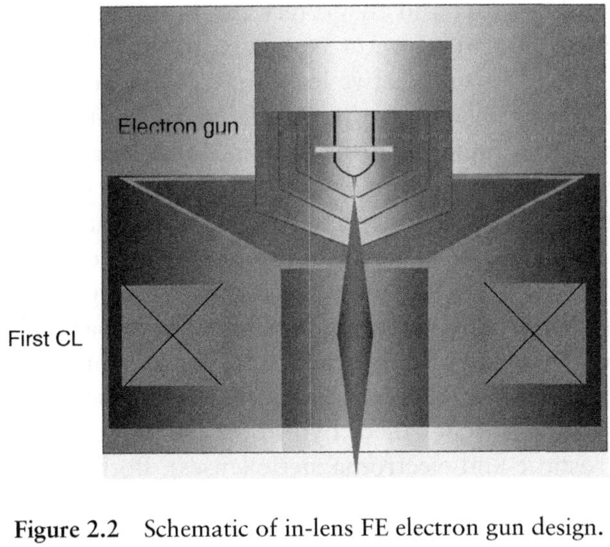

Figure 2.2 Schematic of in-lens FE electron gun design.

current, but it can also maintain a small spot size at the sample. This design improvement translates into the ability to perform microanalysis at lower kVs without sacrificing probe size or probe current (for example it is possible to obtain 5 nm spot in 5 nA beam at 3 kV).

Another major improvement in the electron sources has been the development of a monochromator for FE-SEM. Though cold field emission sources have been attractive for their low energy spread and operation at room temperature as well as superior image resolution (Crewe *et al.*, 1968; Schwind *et al.*, 2006; Pawley, 1997), their inherent instability, vacuum requirements, sensitivity to contamination and relatively small beam current make them somewhat problematic as a general use instruments for both imaging and microanalysis. The new monochromator design for FE-SEM reduces the energy spread of the thermal field emission source to less than 0.2 eV; as a result the image resolution at lower kVs is improved without significant sacrifice to the high current capabilities of the Schottky source. A more detailed discussion of the monochromator design and performance at low voltages is presented in Chapter 3.

2.3 SEM COLUMN DESIGN CONSIDERATIONS

The lens system of the SEM column generates a de-magnified image of the electron source at the sample and ideally should do so without any increase in beam energy spread or loss of brightness. The ability of the operator to adjust probe current continuously over a wide range (from pA to hundreds of nA) and to protect the beam against the effect of external stray fields are important aspects of the overall design. The engineering of the final lens (objective lens) should take into account several considerations: to provide superior image resolution at all beam energies and currents with samples of all sizes, shapes and characteristics (insulators, magnetic samples, etc.); an ability to use a variety of detectors under different operating conditions and to perform microanalysis (EDS, WDS, CL, etc.) at a working distance appropriate for high-resolution imaging.

Two main types of SEM column design have emerged to address the improvements in overall imaging performance with particular attention towards the low voltage microscopy imaging and microanalysis. The main difference between the two design approaches is the use of electromagnetic lenses versus so called ‘hybrid’ or ‘compound’ lenses (combination electrostatic and electromagnetic lenses). Both types of columns feature a number of different electron detectors, most notably some form of an in-lens detector (secondary and/or backscatter), typically combined with an energy filter. Another, relatively uncommon (for SEM) design involves placing the specimen inside a gap of the split objective lens pole piece (true in-lens system, for example, the Hitachi SU9000). Though there are distinct differences in the approaches to obtain the best overall microscope performance in low voltage regime, there are several general trends followed in contemporary column designs:

1. Use of beam deceleration to reduce effects of lens aberrations while maintaining small probe size; either via use of beam booster (compound lens) and/or application of specimen bias.
2. Minimization of the effects of chromatic aberration; via use of monochromator (Chapter 3) or cold field emission gun.
3. Optimization of analytical performance at low voltages, especially ability to deliver high probe current in small spot size at low voltage.
4. Utilization of various signal filtering schemes and detectors to minimize the effects of specimen charging and improve low kV imaging resolution.
5. Development of lens aberration correctors.

The SEM column design incorporating only electromagnetic lenses has been available for decades; an example of such SEM column (JEOL Ltd.) is shown in Figure 2.3. The column features Schottky field emission

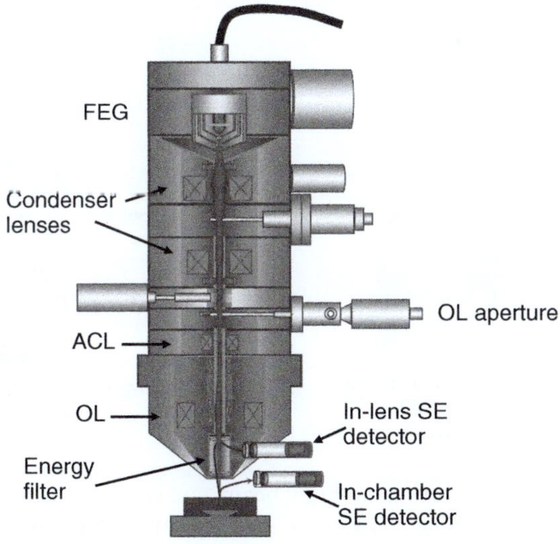

Figure 2.3 Schematic of Schottky FE-SEM immersion lens electron column design (JEOL JSM-7600F).

electron source, a double condenser lens, an electromagnetic objective lens as well as an additional lens, called Aperture Angle Control Lens (ACL). The condenser lens design allows the operator to vary the probe current continuously from several pA to several hundreds of nA, while the ACL automatically optimizes for both high resolution imaging at the lower probe currents, and high spatial resolution X-ray analysis at high probe current. The ACL works by taking into account effects of all aberrations on spot size, and by optimizing the convergence angle accordingly in an automatic fashion. When the SEM is optimized for the smallest probe size (largest conversion angle), some beam scatter (and X-ray generation) away from the immediate probe is possible. For low beam current applications (such as basic high resolution imaging), this is insignificant. When the ACL is optimized for image resolution, the resulting high-current image has a somewhat 'hazy' background, but shows great resolution. The best analytical data comes from the smallest convergence angle; therefore, when the ACL is optimized for analytical work, the ultimate resolution is slightly decreased, yet the analytical signal is no longer affected by the beam tailing, resulting in smaller amount of signal delocalization (see Figure 2.4). The automatic optimization of the ACL maintains high resolution imaging at a wide range of accelerating voltages and probe currents, from a few picoamps to hundreds of nanoamps. For example, the probe resolution at 5 kV at analytical working distance is 2.4 nm at 50 pA, 4.5 nm at 5 nA, and

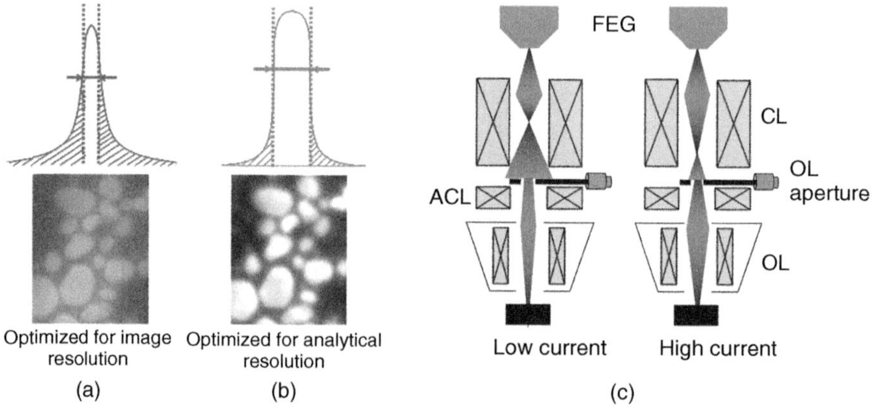

Figure 2.4 ACL optimization at high beam currents for (a) image resolution, and (b) analytical resolution. Au on C images at 15 kV, 20 kX and 198 nA. (c) ACL works by optimizing the convergence angle (α) over a range of accelerating voltages and probe currents to produce the best imaging and analytical instrument performance.

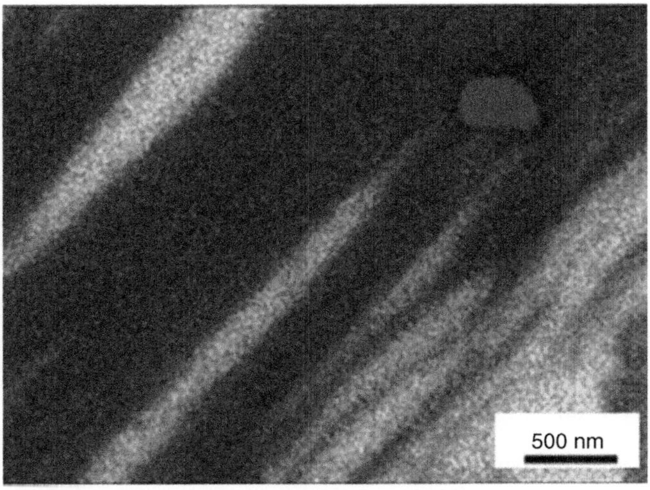

Figure 2.5 EDS map of Sphene $CaTi(Si0_4)(O, OH, F)$, Apatite in a magnesium aluminosilicate matrix with sub-100nm resolution. The map was collected at 8 kV for 20 min using beam current of 20 nA (Courtesy of JEOL USA Inc.) (See plate section for coloured version).

12 nm at 100 nA; this allows collection of very fast EDS/WDS/CL signals with high spatial resolution (sub 100 nm) on bulk samples at low kVs (see Figure 2.5).

The birth of the compound (or hybrid) electron optical column design idea dates back to the days of the Cambridge University SEM group and incorporates a completely different optical principle than the one described before. An example of such a design is a GEMINI® column by Zeiss (see Figure 2.6a; Jaksch and Vermeulen, 2005). As discussed earlier, one of the main considerations for low voltage imaging is beam broadening and chromatic aberration. The energy spread of the beam increases by electron–electron interactions in the crossover regions of the beam pathway (so called Boersch effect) and the resulting mean quadratic energy broadening is the sum of the contributions of the individual crossovers (Boersch, 1954; Rose *et al.*, 1983). In order to avoid potential deterioration of the image resolution due to beam broadening, the GEMINI® column design avoids crossovers. This type of SEM column features a beam booster that maintains high positive potential (e.g., 8 keV) throughout the whole column, regardless of the operating voltage selected by the user. The beam is then decelerated to its final landing energy after passing through the scanning system located in the pole piece

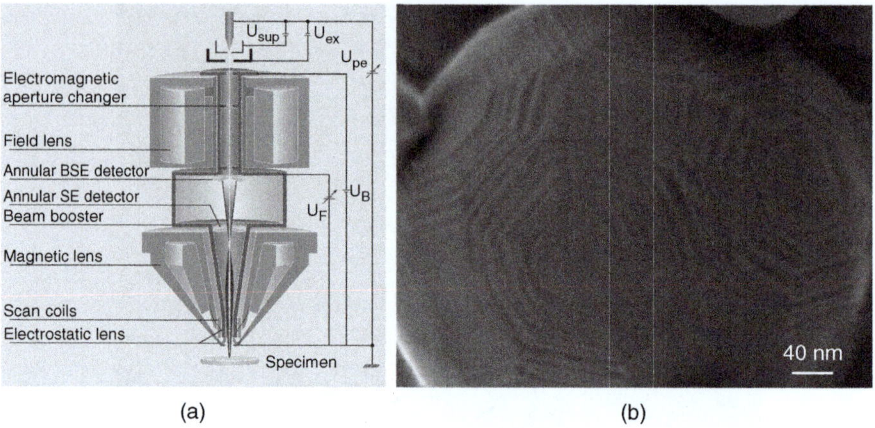

Figure 2.6 (a) GEMINI® column design; (b) high resolution FE-SEM image of alumina spheres taken at 0.45 kV using MERLIN® electron column (courtesy of Carl Zeiss NTS).

gap of the objective lens. The combination electrostatic/electromagnetic lens minimizes magnetic field effects at the sample and is relatively impervious to stray magnetic fields. Since the beam energy E is always maintained at a high value, the transverse chromatic aberration (scales as $\alpha \Delta E/E$, α is the convergence angle) has minimal effect on imaging resolution, particularly at low voltages (Figure 2.6b). The high beam energy and the small length electron pathway minimize beam broadening resulting from stochastic Coulomb interactions between beam electrons. Beam broadening is further reduced by effectively rejecting all excess electrons close to the gun that don't fit into the illuminating beam aperture angle of the objective lens. As a result, there is minimal loss in beam brightness, even at the lowest electron beam energies. The lens is characterized by low values for the spherical and chromatic aberration coefficients, which decrease in proportion with the decreasing accelerating voltage. The adjustment of the probe current, though, is not continuous and is achieved via electromagnetic multi-hole aperture changer incorporated in the upper part of the beam accelerator (Jaksch and Martin, 1995). The beam is effectively deflected through one of the apertures, which determines the convergence angle and the probe current being used. The in-lens signal detection occurs via acceleration of the low energy secondary electrons by the electrostatic lens field and use of an annular detector above the final lens.

The approach employed by the microscope manufacturers in the last few years has been to merge the two important concepts outlined above: combine the high beam current performance for microanalysis with the compound or hybrid objective lens design for the best imaging and analytical performance, particularly at low voltages. A recent example of such system is a TTLS (Through-the-Lens System) design (Asahina *et al.*, 2011) that combines the hybrid lens design with through-the lens detectors and the ACL technology to maintain ultimate analytical performance at high beam currents.

2.4 BEAM DECELERATION

Another approach to reducing the effects of lens aberrations and improvement in low voltage performance is the use of retarding field (or beam deceleration) in front of an intermediate or final cross-over image. This method has also been referred to as 'cathode lens' (Mullerova and Frank, 1993; Frank *et al.*, 2007). The accelerating voltage which along with lens aberrations determines the minimum probe size and thus the resolution limit is retarded by a negatively charged stage bias to a lower landing energy. The landing voltage ($E_{landing} = E_{gun} - E_{bias}$) can be varied with a combination of gun voltage and specimen bias to achieve the necessary imaging performance and obtain high resolution ultra-low kV results, with typical values from 0–5 kV for specimen bias. In fact, beam deceleration serves as a form of aberration correction; the aberration coefficients (both spherical and chromatic) are reduced when the ratio $E_{landing}/E_{gun}$ is decreased for a fixed E_{gun}, meaning higher specimen bias produces enhanced resolution at ultra-low kV (Mullerova and Lenc, 1992; Mullerova and Frank, 1993). This approach preserves the advantages of high kV imaging (gun brightness, small probe size) with added advantages of low landing voltage (reduced charging, specimen contamination, improved surface detail). A practical advantage of this approach is that the influence of the stray magnetic fields on imaging is also significantly reduced, since the low energy primary electrons are only confined to a small area on the sample surface and the gun is operated at higher kV. An additional benefit is that specimen bias and the resulting electric field overcome the small electric fields that may exist at the sample surface under standard imaging conditions, and thus remove the streaking, localized charging and any other disruptions to the emitted electron trajectories. Figure 2.7 shows beam deceleration configuration in a semi in-lens system. When a polymer crystal specimen

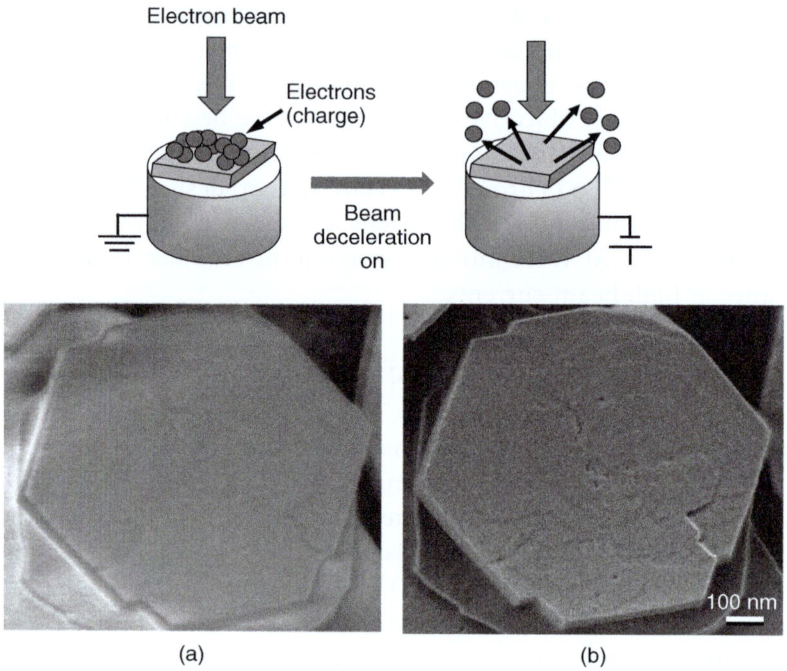

Figure 2.7 Schematic of stage bias used for beam deceleration (top). Example of polymer crystal observed at 0.5 kV before (a) and after (b) beam deceleration (courtesy of JEOL USA).

is imaged without beam deceleration (Figure 2.7, bottom), even at 0.5 kV the primary beam scanning the sample is significantly deflected by the specimen charge and obscures any surface detail; application of specimen bias dramatically improves the imaging conditions and allows high resolution observation of crystal surface. Use of beam deceleration in commercial SEM facilitates specimen observation in ultra-low voltage regime, down to few tens of eV.

Beam deceleration is an excellent approach when the specimen is relatively flat (or for instance comes in a well dispersed powder form); however, it is somewhat problematic when the specimen has large topography variations or is tilted with respect to the incident beam. In such cases, even small movements across the specimen change the characteristics of the localized field created by specimen bias and require continuous operator adjustment of the alignment parameters.

2.5 NOVEL DETECTOR OPTIONS AND ENERGY FILTERS

2.5.1 Secondary Detectors

The ability of the electron source and column to focus the electron beam into the smallest spot size at the sample surface at low voltages is only part of the equation. It is equally important to utilize a detector system that will adequately detect and record the emitted signals. A variety of electron detection approaches have been implemented on the current state of the art microscope systems; we will focus on the new electron filtering capabilities and backscatter detectors specifically designed for low voltage applications.

In order to achieve the highest resolution in the SEM, regardless of the accelerating voltage, the first step is to shorten the focal length of the objective lens (i.e., working distance) and thus place the specimen in the flux field of the immersion lens. While in-lens SE detectors in the immersion lens microscopes are very efficient in collecting electrons, the inherent problem is that the entire range the electrons is collected, including the lowest energy electrons that also carry majority of the charge that can be detrimental for imaging. In order to efficiently filter the signal entering the in-lens secondary detector, two different approaches have been employed for commercial immersion lens microscopes: ExB filter (Hitachi High Technologies) and r-filter (JEOL Ltd.). ExB filter is effectively a Wien filter operated in conjunction with an in-lens detector (Sato *et al.*, 1993; Joens, 2001). The filter consists of a tube, with the electric and magnetic fields perpendicular to the SE trajectory (schematically shown in Figure 2.8a); only the electrons of certain kinetic energy can pass through the filter without being deflected by the corresponding Coulomb and Lorenz forces. The non-filtered electrons collide with the tube walls and are not detected by the in-lens detector. When a positive voltage is applied to the electrode plates, both secondary and backscatter electrons spiral up the column, where backscatter electrons are converted into secondary electrons via ExB filter and are detected by the in-lens detector. This mode produces highly surface sensitive images, but can be also severely prone to charging effects. Application of negative voltage to the electrode plates results in rejection of low energy secondary electrons and the detected signal is dominated by high-energy electrons carrying backscatter image information.

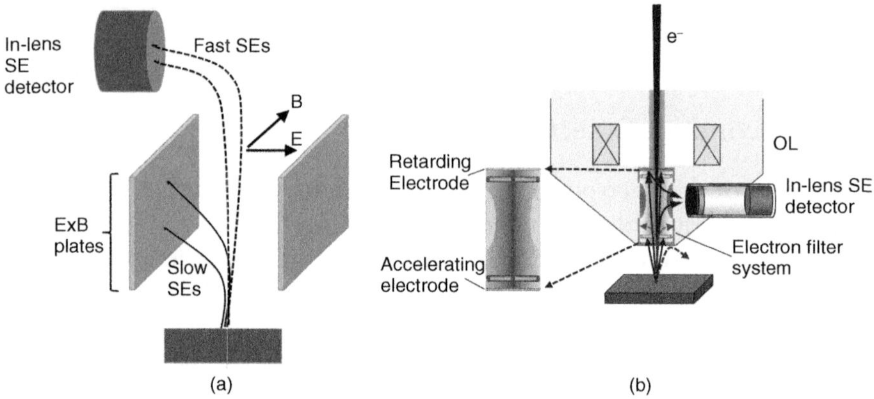

Figure 2.8 Energy filtering in semi-in lens systems. (a) ExB filter (Hitachi) and (b) r-filter (JEOL).

The r-filter schematic (JEOL Ltd) is shown in Figure 2.8(b). The filter components include a cylindrical conversion electrode placed inside the objective lens, with retarding anode and cathode combination for selective detection or elimination of the electrons (Kazumori, 2002). The electric field produced by the electrodes deflects the secondary electron paths, causing very low-energy electrons (charge) to be deflected and eliminated by collision with the cylindrical electrode. The selection of voltage applied to the electrodes allows a continuous adjustment of the signal detected by the in-lens SE detector. This ability to selectively detect different electron energies using the in-lens SE detector via r-filter can effectively give the user capability to image either lower energy secondary component of the signal (SE mode) or a mixture of secondary/backscatter component (Sb or Bs modes) or a higher energy backscatter component (BE mode). Figure 2.9 shows an example of a sample of tin spheres encapsulated in alumina imaged with various modes of the r-filter. This sample is somewhat complex for imaging, since it contains intertwined areas of high-*Z* conductive material (tin) and low-*Z* charging material (alumina). In the SE mode only very low energy electrons are detected, therefore the image carries a substantial amount of charge contribution and appears somewhat 'flat'. In the Sb mode, the lowest energy electrons are suppressed and slightly higher energy electrons are allowed to contribute to the image; Bs image has even stronger contribution of high energy electrons; hence, the tin balls have stronger contrast, similar to backscatter image. In the BE image, low energy electron contribution is completely suppressed, and the image

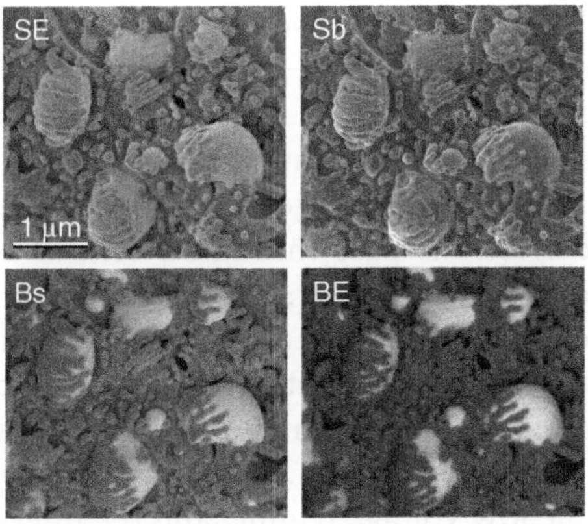

Figure 2.9 Specimen of tin balls covered in Al_2O_3 imaged with different modes of r-filter at 2 kV (courtesy of JEOL USA).

formed with an in-lens secondary detector and an r-filter has effectively a backscatter image contrast. The filtering modes, of course, perform the best when the sample is fairly close to the objective lens (short working distance) to collect the highest amount of signal and obtain the best signal to noise ratio. The use of filters serves also an alternative to backscatter imaging for beam sensitive or highly charging specimens where low voltages and/or high beam currents are inapplicable.

2.5.2 Backscatter Detectors

At low electron voltages the spatial resolution in the backscattered electron image is significantly improved due to the reduced interaction volume. The key for backscatter imaging in the low-voltage regime then becomes the ability to achieve adequate signal to noise ratio, high sensitivity and the ability to effectively provide an equivalent of energy-filtered images. In the past, the backscatter imaging at low voltages was limited to tens of nanometers resolution at best with relatively poor signal to noise ratio. The new BSE detectors are capable of resolution down to several nanometers and can operate in sub-1 kV regime. This makes current BSE detectors not only capable of providing highly surface-specific information, but also a viable imaging option when the specimens are highly charging, since the backscatter signal is substantially less affected by charge compared to secondary electrons.

For hybrid lens systems (combination of electrostatic and electromagnetic lenses) the filtering solution typically comes in the form of a fixed on-axis in-lens scintillator-type detector with a filtering grid in front of it. An example of such a detector is EsB (Energy selective Backscatter) detector in GEMINI® (Zeiss) column (Figure 2.10a); this energy and angle selective backscatter detector is positioned directly above the in-lens SE detector (Steigerwald *et al.*, 2004; Jaksch, 2008). The backscattered electrons pass through a central aperture and then are projected onto the EsB detector; a filtering grid positioned in front of the detector repels any secondary electrons that may have passed through the aperture. The grid voltage can be adjusted from 0 to up to −3 kV, to produce backscatter images with varying electron energy contribution based on the specimen. This detector is particularly useful when working in the low voltage regime, and the filtering grid can select very narrow bands of energy to display various types of information (voltage contrast, surface information, material contrast differences etc.). Figure 2.10(b) shows an example of carbon nanotubes (CNT) with Pt nanoparticles on the surface of CNTs imaged with EsB detector using different grid voltages at 3 kV accelerating voltage. Image collected with no grid bias shows effectively an SE type contrast, with a view of overall CNT morphology and minimal contrast from Pt particles. Bias setting of −400 V produces a 'mixed' signal image with some information about Pt nanoparticle size and location. Further increase in grid bias (−500 V) suppresses SE contrast and shows what looks effectively like BSE image with bright contrast from high atomic number Pt.

Imaging at low voltages also spurred improvements in traditional style backscatter detectors, which typically can be inserted and retracted below the objective lens. These new detectors are often solid-state type with an ultra-shallow doped layer on the surface of the chip for improved sensitivity. One example of such detector is a DBS (Distributed Backscatter Detector) by the FEI Company, an annular eight-segment detector that allows detection of different backscatter emission angles via a specific selection of the segments contributing to the image (Gestmann *et al.*, 2010). Two distinctly different modes of operation—CBS (concentric backscatter) and ABS (angular backscatter)—enable different specimen information to be collected. This detector can also work in conjunction with the specimen bias and immersion lens field to produce different types of material contrast depending on the application.

Another example of a new style backscatter detector is LABe (Low Angle Backscatter electron) detector (JEOL Ltd), a very low profile retractable solid-state detector (Erdman *et al.*, 2009; Schwandt, 2010). The LABe detector can be used in two capacities: (1) at moderate kVs and longer working distances it serves as a traditional backscatter detector,

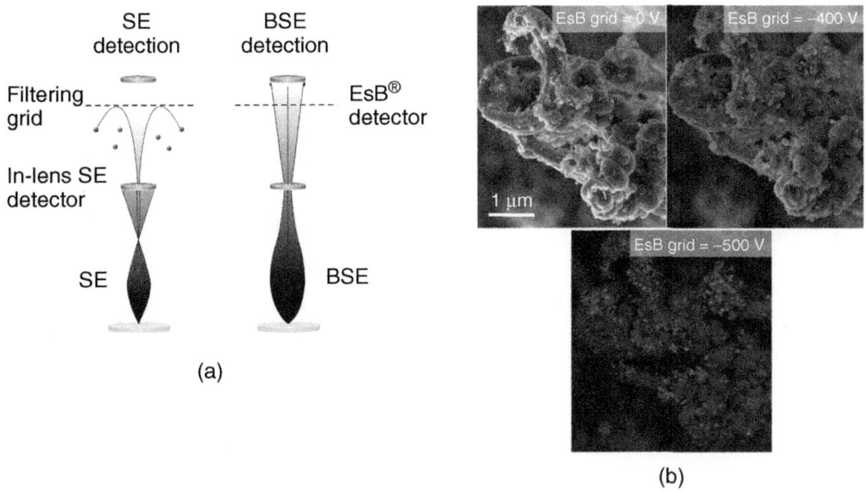

Figure 2.10 (a) EsB detector detector design; (b) example of carbon nanotubes covered with Pt nanoparticles imaged at 3 kV with EsB detector at different filtering grid voltages.

detecting high angle backscatter signal and yielding *Z*-contrast and channeling contrast image; (2) at short working distances (3.5–6mm) it can detect low angle backscatter signal at low voltages, thus providing highly surface sensitive imaging. Low-angle BSEs escape from only the top few nanometers of the surface, although they are generated throughout the entire interaction volume of the specimen; by lowering the accelerating voltage down to a few kVs, or even a few hundred volts, the beam/specimen interaction volume is also significantly reduced. Combining LABe with beam deceleration further enhances its capabilities and acquires images down to few hundred volts with backscatter detector, a particularly useful feature when the sample has a charging component. The choice of accelerating voltage, specimen bias and specimen-detector geometry effectively results in the ability to perform angle-filtered BSE imaging (Figure 2.11a). This yields true surface information as well as crystallographic contrast. An example of the combination of the beam deceleration and the LABe for low kV BSE imaging is shown in Figure 2.11(b). The 1 kV image clearly shows enhanced surface information (including surface contamination) when compared to the 2 kV image. The images acquired with no specimen bias show typical BSE compositional and crystallographic information. An increase in specimen bias (from 0 V to 2 kV) suppresses to some extent the more traditional compositional contrast, and shows enhanced

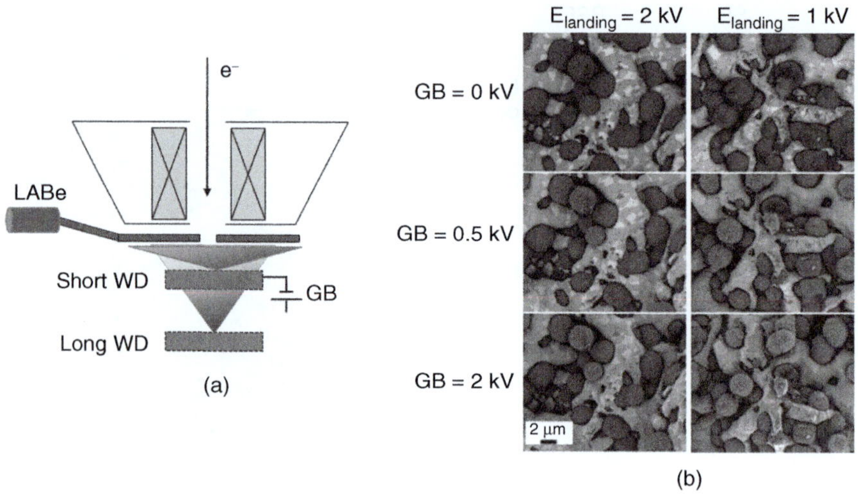

Figure 2.11 (a) Schematic of LABe detector operation; (b) sample of uncoated ceramic eutectic imaged with variations of landing voltage (1 and 2 kV) and specimen bias (GB values) using LABe detector.

surface detail as well as additional morphological information. The combination of specimen bias and voltage chosen dictate the type of information acquired with the LABe detector.

2.6 LOW VOLTAGE STEM IN SEM

Scanning transmission electron microscopy (STEM) has traditionally been performed using either dedicated instruments or detectors at accelerating voltages of 120 to 300 kV. However, similar imaging can be performed inside an SEM by using ultra-thin specimens and collecting the signal transmitted through the specimen. Since SEM operates at accelerating voltage of 30 kV and below, this is considered to be low kV STEM analysis, with typical resolution in this mode in FE-SEM approaching 6 Å. In some SEM instruments, such as in the Hitachi S-5500 (or SU9000) where the sample is located directly inside the objective lens, imaging of carbon lattice fringes has been demonstrated (Ngo *et al.*, 2007). While low kV STEM cannot compete in resolution against dedicated S/TEMs, reduction of the accelerating voltage down to 30 kV and lower there is enhancement of image contrast, especially for low atomic number components and low density materials, making it ideal for the characterization of polymers or biological samples (Golla, 1994). STEM-in-SEM is becoming exceedingly popular with biologists, polymer

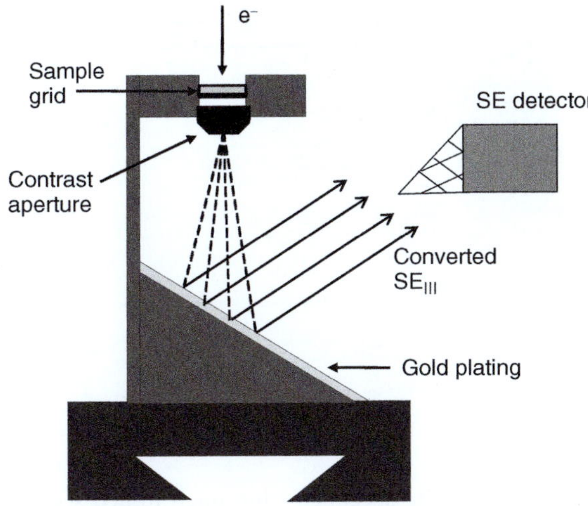

Figure 2.12 STEM in SEM converter holder.

scientists and materials scientists looking at nanoparticles for its ease of use, cost effectiveness and high-resolution. It is imperative, though, to have a very thin specimen for STEM-in-SEM observation, often times thinner than for a typical TEM analysis (see Chapters 4 and 5).

There are two configurations traditionally used for STEM-in-SEM analysis: (1) STEM converter, and (2) dedicated STEM detector. STEM converter is a price-conscious option, shown in Figure 2.12. A STEM converter holds a 3 mm diameter specimen (typical standard specimen size for TEM/STEM holders) and has a hole through it, allowing the beam to transmit through. The transmitted beam of electrons then hits the angled Au-coated mirror surface, where the transmitted electrons are converted into secondary electrons. These secondary electrons are subsequently detected by the in-chamber SE detector. By selecting various 'aperture' sizes below the sample, the user can enhance image contrast; however, the smallest aperture (highest contrast) also means reduced field of view. The STEM converter is an inherently noisier imaging option and only offers bright-field imaging in its standard configuration.

The second option is to use a dedicated STEM detector (scintillator/photomultiplier or solid state type) to collect the transmitted signal in much the same way as in a traditional TEM/STEM. The detector can collect both bright-field signal as well as annular dark-field image. Depending on the detector type, the user can often select different

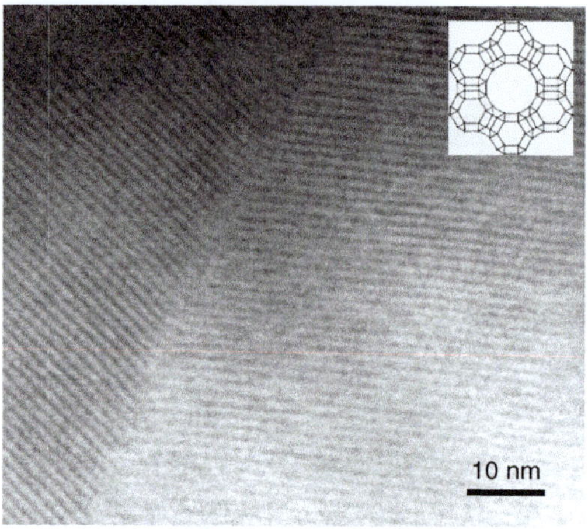

Figure 2.13 Image of faujasite specimen (thin microtomed section) imaged with STEM-in-SEM on JSM-7600F. The fringes correspond to the pore diameter of 7.4 A. Inset: faujasite structure.

collection angles by for example engaging or disengaging various segments of multi-segment solid-state detector, though image interpretation can be somewhat complex. In bright-field mode, on-axis with the electron beam, the image is formed by collecting weakly scattered transmitted electrons; in the dark-field mode electrons scattered at higher angles are collected, and resulting image contrast is associated with the atomic number and mass-thickness contrast. When imaging with low kV STEM beam/specimen interaction volume is substantially reduced compared to a bulk sample, resulting in improved resolution and enhanced contrast, reduced charging, less potential specimen damage and reduced contamination. Figure 2.13 shows an example of high-resolution STEM image of faujasite zeolite. The faujasite framework consists of sodalite cages that are connected through hexagonal prisms. The STEM image shows fringes that correspond to the pore diameter spacing of 7.4 Å (the pores are formed by 12-membered rings).

2.7 ABERRATION CORRECTION IN SEM

In SEM the dominant aberrations that deteriorate image resolution are the first order chromatic and third-order spherical aberrations of the objective lens. Since Scherzer determination in 1936 that in rotationally symmetrical lenses the spherical and chromatic aberration coefficients are always positive, as long as the beam travels in a charge-free region

(Scherzer, 1936), the use of multipole correctors has emerged as a prominent way to overcome these limitations and improve electron microscope resolution (Zach, 1989; Zach and Haider, 1995).

The JEOL JSM-7700F is the first and so far the only commercially available aberration-corrected scanning electron microscope. The JSM-7700F aberration corrected SEM design incorporates a cold field emission source, a multi-pole corrector lens system and corrects via computer control the lens aberrations (both Cc and Cs), which would otherwise limit the ultimate electron probe resolution (Honda and Takashima, 2003; Kawasaki *et al.*, 2005; Kazumori *et al.*, 2004; Uno *et al.*, 2005). The corrector, which is based on Zach's design (Zach and Haider, 1995; Zach, 1989), consists of four sets of 12-pole elements, and is placed between the aperture angle control lens (ACL) and the objective lens (Figure 2.14). For successful and easy operation of the SEM it is necessary to be able to make all the necessary alignment adjustment both with the corrector 'on' on 'off'; though the fact that the electron gun, ACL, OL and the corrector need to be adjusted for multiple accelerating voltages (unlike TEM where discrete and limited set of kVs is employed) presents a design challenge. In the case of JSM-7700F the gun, condenser lens and ACL are adjusted to form the incident beam at the center of the quadrupole component of the first of 12 poles. The appropriate ACL and OL settings are recorded and set to maintain the smallest probe size at the specimen plane. This minimizes the difference in electron rays pathways between corrector 'on' and 'off' setting. Another design hurdle is a

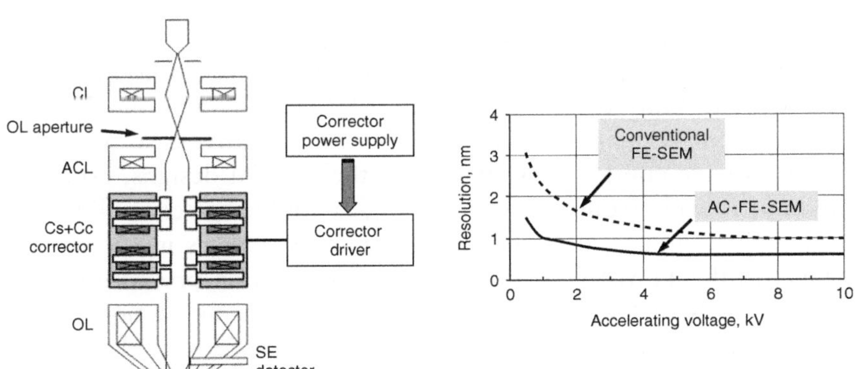

Figure 2.14 JEOL JSM-7700F aberration corrected FE-SEM design (left). Comparison of spot size (resolution) versus accelerating voltage for conventional vs. aberration corrected FE-SEM.

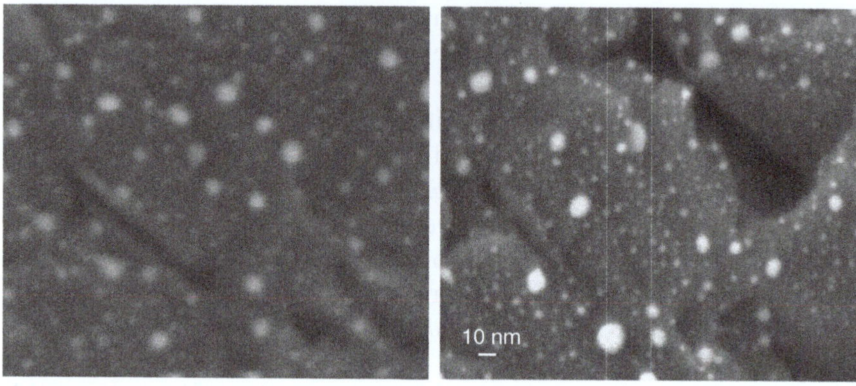

Figure 2.15 Improvement in resolution at low voltage with AC-SEM: images of Au on C at 1 kV without (left) and with (right) aberration corrector (courtesy of JEOL Ltd).

possible ripple in the corrector power supply that could limit aberration correction to higher voltages only. This issue was overcome by selecting a focal plane of the condenser lens closest to the aperture in order to minimize the electric field of the quadrupole; additionally, a low pass filter was used to minimize stray capacitance between the poles. These design improvements produced first of its kind aberration corrected SEM, with 1 nm resolution at 1 kV and 0.6 nm resolution at 5 kV (more than double the resolution of a conventional FE-SEM under similar conditions). Figure 2.15 shows a comparison between images taken with a conventional FE-SEM and an aberration-corrected FE-SEM.

Though advantages of using aberration correction in SEM to improve instrument resolution are undisputed, it is important to consider possible drawbacks. The biggest potential problem may be versatility: the users operating the SEM expect to be able to change conditions on a fly going between very high and very low kV with relative ease. While the correctors have come a long way from their initial designs, especially with improved electronics and computing capabilities, it can still be time consuming to adjust correctors for substantial changes in accelerating voltage. In the TEM/STEM the aberration correction procedure uses a Ronchigram (Cowley, 1986)—a type of defocused diffraction pattern from electron transparent sample—to precisely correct aberrations. In the SEM no such specimen/procedure are available, therefore the correction process is much more convoluted. Uno *et al.* (2005) proposed a method of automated aberration correction via precise digitizing of the amplitudes of aberrations through analysis of the under and over focused probe shapes extracted from SEM images. This procedure can automatically correct all axial geometrical aberrations up to the third order by a linear feedback control method.

Some of the other issues with using aberration correctors in SEM involve loss of depth of field (Goldstein *et al.*, 2003; Joy, 2008) and the limitations of the specimen shape and size suitable for aberration correction procedure. These become particularly important in metrology applications (Sato and Mizuno, 2000).

For most users, though, the biggest issue with aberration corrected SEM will probably be the cost involved not only in purchase but also maintenance of such system (the addition of aberration corrector likely places the instrument at over $1M price tag). While the benefits of aberration correction especially at low voltages are significant, the optimal system will likely have the capability to turn the corrector 'on' or 'off' seamlessly such that the overall system is more versatile and user-friendly.

2.8 CONCLUSIONS

Low voltage SEM development has progressed greatly over the last 10 years with the ability to operate the accelerating voltage of the SEM down to the tens of volts, this being truly low voltage. The key area of development is the reduction of lens aberrations at lower voltages using a number of techniques from column design to beam landing energy adjustment and spherical/chromatic aberration correctors. A key advance in low voltage SEM is the improvement of the electron detectors both the secondary and backscattered detection, since the efficient use of every electron to generate and record signals is critical. The improvement of scanning electron microscopes design continues with emphasis on low voltage capability, low voltage resolution as well as automation of alignment procedures and new detector capabilities.

REFERENCES

S. Asahina, S. Uno, M. Suga, *et al.*, A new HRSEM approach to observe fine structures of novel nanostructured materials, *Micropor. Mesopor. Mater.*, **146**, 11–17 (2011).

H. Boersch, Experimentale bestimmung der energieverteilung in thermisch ausgleoesten elektronenstrahlen. *Z. Phys.*, **139**, 115 (1954).

A.N. Broers, Recent advances in SEM with lanthanum hexaboride cathodes, *Scanning Electr. Microsc.*, 10–18 (1974).

J.M. Cowley, Electron diffraction phenomena observed with a high resolution STEM instrument, *J. Electr. Microsc. Tech.*, **3**, 25–28 (1986).

A.V. Crewe, D.N. Eggenberger, J. Wall, and L.M. Welter, Electron gun using field emission source, *Rev. Sci. Instr.*, **39**, 576–583 (1968).

N. Erdman, N. Kikuchi, V. Robertson, and T. Laudate, Multispectral imaging in an FEG-SEM, *Adv. Mat. Proc.*, **9**, 28–31 (2009).

L. Frank, I. Mullerova, K. Matsuda, I. Susumu, Cathode lens mode of the SEM in materials science, *Mater. Transact.*, **48**, 944–948 (2007).

I. Gestmann, K. Kooijman, A. Sakic, *et al.*, New solid state detector design for ultra-sensitive back-scattered electron detection, *17th International Microscopy Congress (IMC17)*, Rio de Janeiro, Brazil (2010).

U. Golla, B. Schindler, and L. Reimer, Contrast in the transmission mode of a low-voltage scanning electron microscope, *J. Microsc.*, **173** (3), 219–225 (1994).

J. Goldstein, D.E. Newbury, D.C. Joy, *et al.*, *Scanning electron microscopy and x-ray microanalysis*, 3rd edn, Springer, New York (2003).

K. Honda, and S. Takashima, Chromatic and spherical aberration correction in the LSI inspection scanning electron microscope, *JEOL News*, **38** (1), 36–40 (2003).

H. Jaksch, Low loss BSE imaging with EsB detection system on the Gemini Ultra FE-SEM, *EMC* 2008, 1, 555–556 (**2008**).

H. Jaksch, and J.P. Martin, High-resolution, low-voltage SEM for true surface imaging and analysis, *Fresenius J. Anal. Chem.*, **353**, 378–382 (1995).

H. Jaksch, and J.P. Vermeulen, New developments in GEMINI® FESEM technology, *Microsc. Today*, **13**, 8–10 (2005).

S. Joens, Hitachi S-4700 ExB filter design and applications, *Microsc. Microanal.*, 7, 878–879 (2001).

D.C. Joy, The aberration corrected SEM, in *Biological Low-Voltage Scanning Electron Microscopy*, H. Schatten and J. Pawley (eds), Springer, New York (2008).

T. Kawasaki, N. Tomonori, and H. Kotoko, Developing an aberration-corrected Schottky emission SEM and method for measuring aberration, *Microelectr. Eng.* **86**, 1017–1020 (2009).

H. Kazumori, Development of JSM-7400F: new secondary electron detection systems permit observation of non-conductive materials, *JEOL News*, **37E** (1), 44–47 (2002).

H. Kazumori, K. Honda, M. Matsuya, and M. Date, Field emission SEM with a spherical and chromatic aberration corrector, in *Proc. 8th Asian Pacific Conf. on Electr. Microsc.*, 52–53 (2004).

D.B. Langmuir, Theoretical limitations of cathode ray tubes, *Proc. Inst. Radio Eng.*, **25** (8), 977–991 (1937).

I. Mullerova and J. Fiser, Magnetic shielding of stray magnetic fields in the scanning electron microscope, *Rev. Sci. Instrum.*, **65**, 1968–1972 (1994).

I. Mullerova, and L. Frank, Very low energy microscopy in commercial SEMs, *Scanning*, **15**, 193–201 (1993).

I. Mullerova, and M. Lenc, Some approaches to low-voltage scanning electron microscopy, *Ultramicroscopy*, **41**, 399–410 (1992).

V.V. Ngo, M. Hernandez, B. Roth, and D.C. Joy, STEM imaging of lattice fringes and beyond in a UHR in-lens field-emission SEM, *Microsc. Today*, **15** (2), 12–16 (2007).

C.W. Oatley, W.C. Nixon, and R.F.W Pease, Scanning Electron Microscopy, *Adv. Electronics and Electron Phys.*, **21**, 181–247 (1965).

J. Pawley, The Development of Field-Emission Scanning Electron Microscopy for Imaging Biological Surfaces, *Scanning*, **19**, 324–336 (1997).

H. Rose, and R. Spehr, Energy broadening in high-density electron and ion beams: The Boersch effect, *Adv. in Electronics and Electron Phys.*, **13 C**, 475–530 (1983).

M. Sato and F. Mizuno, Depth of field at high magnifications of scanning electron microscopes, *J. Vac. Sci. Technol. B*, **18**, 3047–3051 (2000).

M. Sato, H. Todokoro, and K. Kageyama, Snorkel-type conical objective lens with E cross B field for detecting secondary electrons, *Proc. SPIE—Charged Particle Optics*, 2014, 17–23 (1993).

O. Scherzer, Über einige Fehler von Elektronenlinsen, *Z. Phys.* **101**, 593 (1936).

C.S. Schwandt, Characterizing nanometer-scale materials using a low-angle backscattered electron detector, *Amer. Lab.*, Nov.15, 13–17 (2010).

G.A. Schwind, G. Magera, and L.W. Swanson, Comparison of parameters for Schottky and cold field emission sources, *J. Vac. Sci. Technol. B*, **24**, 2897–2901 (2006).

M.D.G. Steigerwald, R. Arnold, J. Bihr, *et al.*, New Detection System for GEMINI, *Microsc. Microanal.*, **10**, 1372–1373 (2004).

L.W. Swanson, and D. Tuggle, Recent progress in thermal field electron source performance, *Appl. Surf. Sci.*, **8**, 185–196 (1981).

R.F.M. Thornley, Recent developments in scanning electron microscopy, in *The Proceedings of the European Regional Conference on Electron Microscopy*, Delft, 173–176 (1960).

R.F.M. Thornley and L. Cartz, Direct examination of ceramic surfaces with the scanning electron microscope, *J. Am. Ceram. Soc.*, **45**, 425–428 (1962).

D.W. Tuggle, and L.W. Swanson, Emission characteristics of the ZrO/W thermal field electron source, *J. Vac. Sci. Tech.*, **B3**, 220–223 (1985).

S. Uno, K. Honda, N. Nakamura, *et al.*, Aberration correction and its automatic control in scanning electron microscopes, *Optik*, **116**, 438–448 (2005).

A. Vladar and M. Postek, The scanning electron microscope, in *Handbook of charged particle optics, 2nd edn*, J. Orloff (ed.), CRC Press, Boca Raton, FL (2008).

J. Zach, Design of a high resolution low voltage scanning electron microscope, *Optik*, **83** (1), 30–40 (1989).

J. Zach, and M. Haider, Aberration correction in a low voltage SEM by a multipole corrector, *Nucl. Instr. Meth. Phys. Res. A*, **363**, 316–325 (1995).

3

Extreme High-Resolution (XHR) SEM Using a Beam Monochromator

Richard J. Young[1], Gerard N.A. van Veen[2], Alexander Henstra[2] and Lubomir Tuma[3]
[1]*FEI Company, USA*
[2]*FEI Company, The Netherlands*
[3]*FEI Company, Czech Republic*

3.1 INTRODUCTION

The scanning electron microscope (SEM) continues to play an essential role in the imaging, characterization and metrology of nanoscale materials and structures. While the SEM does not have the ultimate imaging resolution and analytical capabilities of the transmission electron microscope (TEM), it nevertheless offers benefits in terms of ease of sample preparation and the ability to image surface details and three-dimensional structures, especially at low beam voltages where reduced electron range and interaction volume provides highly surface sensitive information [1, 2].

Major challenges for extending the performance of the low voltage SEM are the inherent aberrations that limit the beam resolution. This chapter reviews these limitations and discusses developments that have helped extend SEM into an extreme high-resolution (XHR) regime, where sub-nanometer resolution can be achieved at low beam voltages. In particular, it will describe the implementation of a source monochromator, which reduces the effects of chromatic aberrations

Low Voltage Electron Microscopy: Principles and Applications, First Edition.
Edited by David C. Bell and Natasha Erdman.

at low voltage, resulting in a more tightly focused electron beam. This capability, first available in the Magellan™ XHR SEM in 2008 [3–6], but since extended to other FEI microscopes (including dual-beam systems combining SEM with focused ion beam: FIB), provides significant benefit for surface sensitive imaging.

3.2 LIMITATIONS IN LOW VOLTAGE SEM PERFORMANCE

Although SEM is considered a surface imaging technique, it would be more accurately called near-surface since a significant portion of the detected signal may originate from hundreds, or even, thousands of nanometers below the surface, depending upon the beam energy used. In many cases, the resolution of the SEM is controlled by the size of this excitation volume, which is determined by the size of the focused probe and the distance beam electrons penetrate into the sample as they spend their energy in signal-generating interactions with local atoms. Beam penetration is easily reduced by decreasing beam energy, but in a conventional SEM the gain from reduced penetration is negated by increased spot size at beam energies below a few keV. In the limit of small beam current, the beam probe size can be approximated as: [7]

$$d50 = \left[\left(0.34 \cdot C_c \alpha \, \Delta U / U\right)^2 + \left(\left(0.54 \cdot \lambda / \alpha\right)^4 + \left(0.18 \cdot C_s \alpha^3\right)^4 \right)^{1/2} \right]^{1/2} . \tag{3.1}$$

where $d50$ is probe size containing 50% of the current, α is semi-angle of beam convergence at the specimen, C_C and C_S are respectively chromatic and spherical aberration coefficients, U is landing voltage, $e\Delta U$ is energy spread of electrons, and λ is electron wavelength. In Equation 3.1 the terms represent, respectively, chromatic aberration, aperture diffraction, and spherical aberration. At higher energies, the spherical aberration is typically the limiting factor, but at low energy it is the chromatic term that becomes the critical one. This is due to the energy spread becoming a larger fraction of the beam energy. This effect can be reduced by decreasing α, but this will increase the diffraction effect.

3.2.1 Aberration Correction

A direct approach to improving performance is to attempt to reduce, or ideally eliminate, the chromatic and spherical aberrations coefficients as happens routinely with light optics. Early in the life of electron microscopy Scherzer proposed that non-cylindrically symmetric lenses could be used to counteract the aberrations of round electronoptical lenses [8]. It took until the last decade for such devices to become commercially available, initially on TEMs. On the SEM, various groups have researched the use of corrector technology, including CEOS (Germany), [9] Philips Research / FEI (Netherlands), [10] Hitachi (Japan) [11] and JEOL (Japan) [12]. However, the perception in the market is that the corrector technology for SEM is still too difficult to handle, despite attempts to automate the corrections, [11, 13] and that it lacks the operational flexibility that is required for day-to-day operation of a SEM.

3.2.2 Electron Source Energy Spread

Before going to the complexity of correctors it seems better to explore alternatives to make further progress in SEM resolution. As the chromatic aberrations are dominant at low energy, the logical next step is to reduce the energy spread in the beam itself. In this respect the cold field emitter (CFE) sources have an advantage to Schottky emitters. An optimized Schottky emitter will in practical situations have a beam spread in the range 0.6–1.0 eV, depending on design details and the resulting Coulomb interactions in the beam. A CFE gives a beam with an energy width in the range 0.25–0.4 eV (with 0.3 eV being typical in SEM operation). However, the Schottky emitter is more stable for both long term drift and short-term statistical beam noise, and provides a greater maximum beam current [14]. These factors are especially important for systems that need to serve multiple applications (including sample material analysis). For dedicated high resolution SEM systems the CFE used to be the preferred choice, but with a beam monochromator it is possible to get all the advantages of the Schottky emitter, with a beam energy spread, and hence low-voltage resolution, that considerably surpasses the performance of CFE.

3.3 BEAM MONOCHROMATOR DESIGN AND IMPLEMENTATION

In order to illustrate the potential of a monochromator for low voltage SEM Figure 3.1(a) shows the probe size as a function of energy spread for several accelerating voltages between 350 V and 2 kV. The theoretical probe size is reported in terms of the 25–75% beam edge resolution based on wave-optical calculations. This value is typically

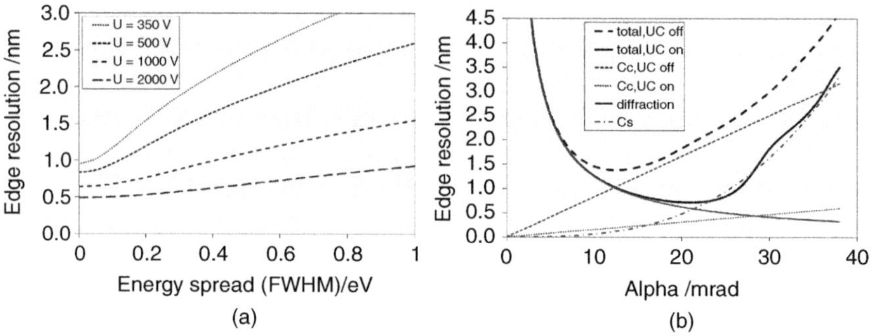

Figure 3.1 Theoretical beam probe size as a function of energy spread (a) and opening angle (b) for Magellan XHR SEM at working distance of 1 mm. The data was determined using wave-optical calculations in the limit of low probe current (i.e., neglecting geometric beam size). Graph (b) shows aberration contributions and resulting probe size at 1 kV for a typical Schottky energy spread (0.8 eV, UC-off) and that obtained with the monochromator (0.15 eV, UC-on). Both parts adapted from Young *et al.*, XHR SEM: enabling extreme high resolution scanning electron microscopy, *Proc. SPIE*, **7378**, 737803 (2009).

$0.6 \cdot d50$. The beam monochromator developed for Magellan enables a reduction of the energy spread to 0.15 eV (FWHM). This value is based both on the theoretical resolution of the monochromator and on actual measurement results using a biased Faraday cup, which have been found to be in good agreement.

To understand more clearly how the resolution improvement is obtained refer to Figure 3.1(b), where the aberrations and resulting probe size, with and without the monochromator, are shown for 1 kV. The situation without the monochromator shows that the minimum point is a balance between the chromatic aberration and diffraction contributions. The reduction in energy spread with the monochromator allows a new minimum point, at a larger optimized aperture semi-angle α, limited more by spherical than chromatic aberration. This increase in α is allowed because, now that chromatic aberrations are lower, you can open up the aperture angle and get a reduced diffraction effect. The result is a smaller probe while maintaining beam current. Cc–Cs correctors have a similar benefit of allowing a larger α while improving resolution.

The monochromated beam for XHR imaging is provided by an electrostatic Schottky-FEG module, which can also produce large probe currents for analysis, shown schematically in Figure 3.2. The extractor plate contains two apertures, producing an on-axis and an off-axis beam. An electrode below the extractor serves as an electrostatic lens.

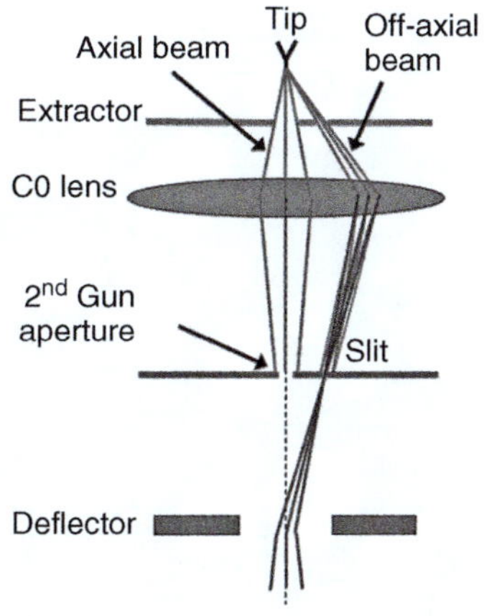

Figure 3.2 Operation of the UC monochromator. For clarity the axial beam is not shown beyond the second aperture. See text for more details (Source: FEI).

This electrode is segmented, so it can also be used for deflection and stigmation. Below this a second aperture plane contains an axial aperture and a small, off-axis slit [15]. The latter is used to monochromate the off-axis beam. A deflector below the second aperture is used to direct either the on-axis or off-axis beam into the rest of the electron column, where a final beam limiting aperture is used to determine the probe current. Unwanted electrons are blocked by apertures either in the gun module, or downstream in the main part of the electron column.

The source module has three modes of operation. In the first, the on-axis beam is used with gun lens off. The second aperture limits the beam current providing low to medium currents (e.g., picoamp to nanoamp currents) for high resolution imaging. In the second mode, the gun lens is turned on to deliver more beam current, which can go to >25 nA (e.g., for analytical applications). The third mode (called UC mode) uses the off-axis monochromated beam. Voltages on the gun lens segments are used to position, focus, and stigmate the off-axis beam on the energy selecting slit. The beam is dispersed due to the off-axial traversal through the strong gun lens, allowing the slit to select just a narrow range of electron energies. The spherical aberration of the gun lens results in coma in the off-axial beam. However, the effect of this coma on the energy spread is negligible for useful probe currents, that is, <100 pA. For larger probe currents, the loss of brightness becomes more important than the reduction in energy spread.

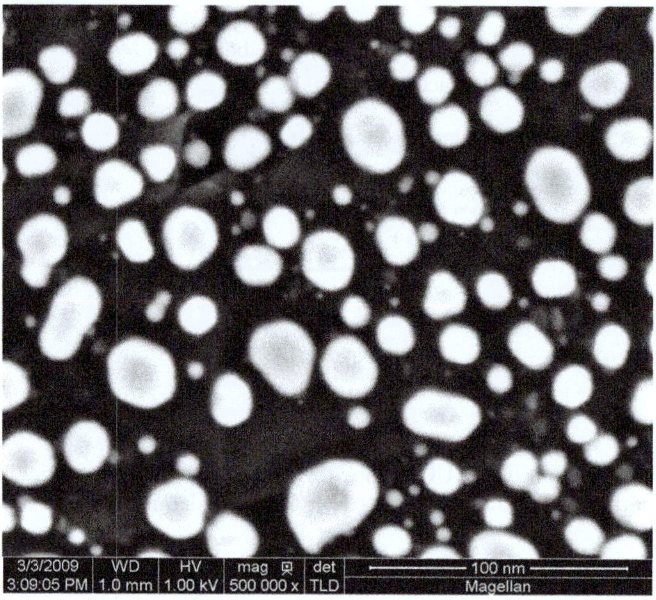

Figure 3.3 Gold-on-carbon image at 1 kV beam voltage using the monochromator (without beam deceleration). Field of view is 256 nm. Adapted from Young *et al.*, XHR SEM: enabling extreme high resolution scanning electron microscopy, *Proc. SPIE*, 7378, 737803 (2009).

The improvement in resolution with the monochromator has been demonstrated to match the theoretical value of 1.8 times improvement at 1 kV, [4] with Figure 3.3 showing an example Au–C image at 1 kV using the monochromator. Image resolution was assessed using the derivative method to measure image sharpness [16, 17]. This measurement determines the width of many thousands of edges in all directions within the image. The average value of the edge width distribution is directly related to the image sharpness, which can subsequently be converted into the edge resolution. The distribution of such values is shown on the left side of Figure 3.4. The beam is clearly well shaped in all directions (no astigmatism). The distribution of values is due to signal-to-noise ratio and to non-idealities in the sample (i.e., the sample is not made up of perfectly sharp edges). On the right of Figure 3.4 the simulated results for an ideal sample (with perfectly sharp edges) imaged with the same theoretical beam point spread function (PSF) and signal-to-noise are shown. Here the data are much less spread, with an average value of 0.73 nm (using 25–75% edge resolution from the theoretical

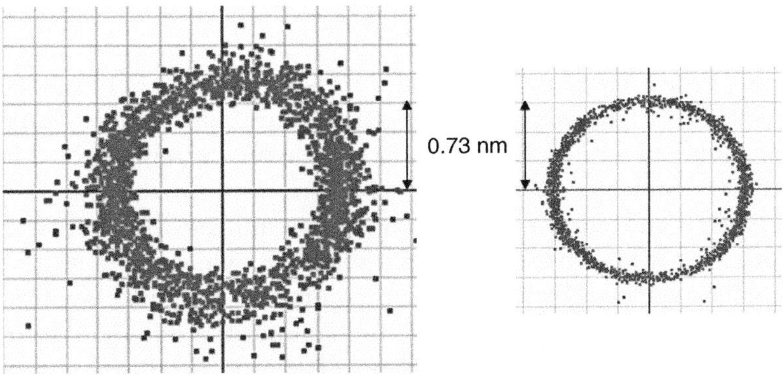

Figure 3.4 Distribution of edge width measurements from an actual 1 kV image (Figure 3.3) on the left and a simulated image on the right. The simulated image uses the theoretical PSF and same signal-to-noise ratio. Adapted from Young *et al.*, XHR SEM: enabling extreme high resolution scanning electron microscopy, *Proc. SPIE*, 7378, 737803 (2009).

beam probe size). This value corresponds to the inner part of the actual image spread, indicating we are essentially demonstrating the theoretical beam performance, with additional spread to larger measured values due to the non-ideality of the sample. Assessing the beam performance using 35–65% edge conversion from image sharpness has been found to be a useful practical resolution criterion taking into account the complete image area and typical sample edge distributions.

3.4 XHR SYSTEMS AND APPLICATIONS

The monochromator module was initially deployed in the FEI Magellan XHR SEM, [3–6] using the Elstar magnetic-immersion electron column, and then subsequently incorporated into the Helios NanoLab™ ×50 family of dual-beam SEM-FIB systems that also uses the Elstar column. A key benefit of the monochromator on the dual-beam type of system is the performance improvement seen at the longer (4 mm) working distance position where the electron and ion beam are coincident; this is important as this is the usual working position for effective dual-beam operation. With the monochromator the 1 kV resolution performance at the 4 mm coincident point equals that of the non-monochromator version of the dual-beam at short (1–2 mm) working distance. Beyond the Elstar column, the monochromator has also been demonstrated to improve the performance of non-immersion lens columns as well.

3.4.1 Elstar XHR Electron Column

The optics and alignment settings of the Elstar column are fully software controlled, and lens and deflection settings are automatically adjusted for optimum resolution at each beam energy and current selected by the user. The beam aperture control mechanism is motorized, and for each beam energy and current the optimum size and position of the beam aperture are set automatically by the software, while also adjusting the various lens settings throughout the column to minimize aberrations. This methodology has benefits for the standard Elstar operation, but is even more critical when the monochromator is added so that the improvement due to the reduced chromatic aberration is maximized. There are no mechanical column alignments and all available alignments have been automated, making their execution much faster than with manual control, and enabling the user to run alignments unattended.

To take maximum advantage of the excellent resolution performance the column should also have very stable behavior. This is because the possibility of using low beam energies and the significantly higher magnifications made feasible by the better resolution, will cause the imaging to be more sensitive to drift and environmental conditions. One of the components for achieving maximum stability is the use of constant-power coils for the magnetic lenses. With constant-power coils the magnetic field of the lens is formed by the summed effect of two individual coils: the currents in these coils are set such that the total power dissipated by the two coils always stays constant, even if the lens field is zero. Constant power technology enables fast switching from high beam voltage to low beam voltage, or vice versa, without any thermal effects, and thus significantly decreases column drift.

The thermal stability in the Elstar column is further enhanced by the use of electrostatic scan and deflection, because this does not generate any power dissipation in the column, as opposed to magnetic scan coils. In addition the linearity, reproducibility and scan speed are much better with electrostatic scanning, due to the absence of hysteresis effects, making it an ideal electron column for lithography and nano-prototyping applications. Much emphasis in the design was put on shielding from the environment. Magnetic shielding performance for both DC and AC fields has been improved significantly by applying double magnetic shielding layers over the whole column length.

Beyond the actual probe size and stability, the image acquisition and image display also need to be considered. For an SEM both the secondary electron (SE) and the back-scattered electron (BSE) detection are important. However, the most relevant quality of an image is its contrast, in particular sufficient contrast to observe the features and information of interest. To get the best performance, complete 3D modeling of the detection section of the Elstar through-lens detector (TLD) was carried out [18]. This modeling includes the electromagnetic fields of the final lens and detectors as well as the SE/BSE creation processes within the sample itself. These simulations allow the tracking of individual emitted electrons based on their energy and angle from the sample into the detectors, and even include the generation and collection of SE3-type electrons (i.e., those created when an SE or BSE from the sample hits part of the column). Modeling shows that the TLD can collect essentially 100% of the SE electrons below 10 eV to allow for low-dose, high-contrast imaging. The modeling approach also has benefits in understanding the optimum detector settings for a particular set of beam conditions, and has already proven very valuable during new detector developments by allowing the optimization of signal collection even before any hardware was constructed or tested.

3.4.2 Beam Deceleration for Extending Low-Voltage Performance

To be able to image at extremely low electron landing voltages, beam deceleration can be applied [2, 19]. In this mode, the stage is negatively biased by potentials of up to 4 kV. This combination of a magnetic and an electrostatic lens gives a further reduction in the probe size. The stage bias also accelerates the SE and BSE away from the sample, providing improved collection by solid-state detectors that would otherwise be insensitive to any BSE leaving the sample at the landing energy itself, and providing additional options for optimizing the contrast within an image.

Figure 3.5 shows the benefit of combining the UC monochromator and beam deceleration. Starting at 30 kV the theoretical probe size is below 0.4 nm. As the beam voltage is reduced, the probe size starts to increase slightly as 10 kV is approached. Moving below 5 kV the non-UC diameter starts to rise more sharply, but switching over to UC mode restores the probe size performance. Similarly in the range 1 kV and below further benefit is obtained by using beam deceleration mode, although UC-mode alone still provides a significant improvement over the standard performance with a factor of more than two at 500 V. Even below 100 V the combination of UC+BD provides remarkable results with a theoretical probe size of one nanometer at 50 V using 4 kV of stage bias.

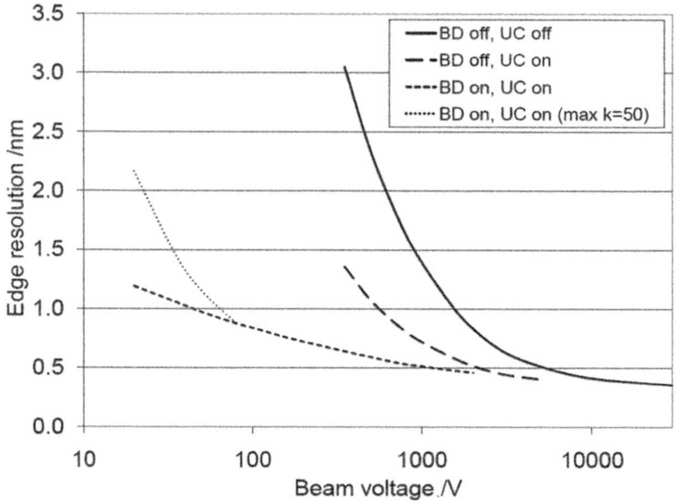

Figure 3.5 Theoretical probe size against beam voltage, with and without UC (monochromator) and BD (beam deceleration). For this data the smallest working distance possible was used, with a minimum value of 1 mm. With beam deceleration the main line uses 4 kV of stage bias, while for the upper line the stage bias is limited to the smaller of 4 kV or 50 times the landing voltage (i.e., 1 kV at 20 V landing energy). Adapted from Young *et al.*, XHR SEM: enabling extreme high resolution scanning electron microscopy, *Proc. SPIE*, **7378**, 737803 (2009).

As the beam energy drops below 100 eV, the distinction between emitted SE and BSE gets blurred as the incoming electrons are similar in energy to the SE themselves. By convention the division between BSE and SEs is a threshold energy of 50 eV, but as Fitting *et al.* discussed, there will be some SE contribution into the >50 eV 'BSE' designation and some BSE that are below 50 eV [20]. Müllerová and Frank [19] describe the various energy regimes in some detail as you move below a few keV and approach zero landing energy, with reference to their considerable experience with beam deceleration. Throughout this range many interesting imaging and contrast phenomenon are present, for example below 100 eV contrast inversions are often seen and they report increased influence of the overall crystal potential and phenomena connected with the electronic structure of the sample. This interesting range of very low beam energies will undoubtedly see further investigation as more systems equipped with beam deceleration are deployed by the various microscope manufacturers.

In addition there may be much to learn from other techniques that also image the sample using electrons in the 0–100 eV landing energy range.

One example is the low energy electron microscope (LEEM), which, while it differs from SEM in being a non-scanning technique, also uses beam deceleration to produce low landing energies. Aberration correction has also been applied to LEEM, which has improved performance by a factor of 2 down to 2 nm [21].

3.4.3 Combination of a Monochromator with Non-Immersion Lens

In principle, the monochromator improves resolution of any system limited by chromatic aberration. There are a number of use cases where use of the immersion lenses (both magnetic and/or beam deceleration) is not optimal. It could be restricted by various technical limitations (special accessories, magnetic samples, sample tilt, low vacuum around the sample, or long working distance), but also complexity, ease of use, and price of the system. For such cases the monochromator can be combined with conventional non-immersion objective lenses. A conventional system with standard Schottky emitter equipped with a monochromator shows resolution improvement by a factor of approximately two at low beam energies (see Figure 3.6) and become comparable, at least from a beam diameter point of view, with immersion systems without a monochromator.

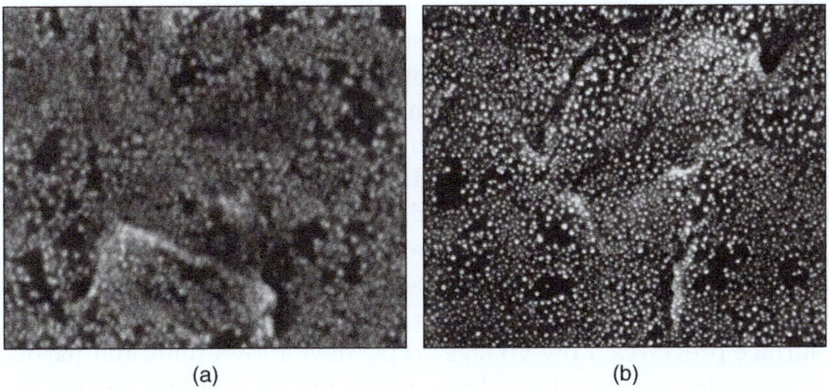

Figure 3.6 Comparison of non-immersion lens performance without (a) and with (b) a beam monochromator (Au-C sample). The 1 kV results are improved by a factor of two from 2.5 nm to 1.2 nm. Each image is 850 nm wide (Source: FEI).

3.4.4 XHR Applications

Extending SEM performance from the ultra-high resolution (UHR) to the XHR regime has the biggest benefit for surface sensitive imaging from about 3 kV and below. To illustrate this we will mention example applications from three diverse disciplines; semiconductor manufacturing, materials research and life sciences.

One common trend across applications is that hydrocarbon compounds from the sample (and if appropriate precautions are not taken, the sample mounting stub, stage and even the system) can cause contamination effects which can severely limit imaging performance at very low kV. Therefore in-situ sample and chamber plasma cleaners are routinely used to reduce contamination effects and to be able to control the optimum sample imaging conditions. The plasma cleaner enables the overall system to be kept clean and also allows direct cleaning of samples already in the system, if this is needed. In addition, liquid nitrogen cooled cryo-traps are also valuable to help maintain a clean imaging environment. The two techniques together have proved highly valuable in obtaining excellent imaging results on samples that were difficult or impossible to work with in systems without these controls.

For semiconductor manufacturers SEM imaging has long been important, but with smaller dimensions the ability of the SEM to see the required details has become more challenging, resulting in more samples being imaged in the TEM. This involves much more involved (and expensive) sample preparation than the SEM route, so XHR imaging has been important in extending the use the SEM for additional node shrinks (see Figure 3.7a). Another semiconductor application that benefits from low energy beams are lithographic resists. In addition to being insulating, and so suffering from beam charging effects, such resists can be highly sensitive to beam exposure, for example, shrinking when exposed to just one or two scans at 1 kV. With such a sample, going to a beam landing voltage of 500 V or below in the XHR SEM produced excellent results with minimal beam effects and good signal-to-noise [4, 22].

Just as semiconductor manufacturers are well into the nanotechnology realm, the research community is also driven by a need for new ways to investigate nanoscale materials, interfaces between materials, and details of surface processes. Low-voltage XHR SEM allows films and particles to be inspected without the addition of conductive coatings, and has been applied to studies of FePt nanoparticles for bit patterned media; [23] iron oxide nanoparticles and films; [24, 25] silicon carbide nanopowers; [26] and fullerene-anion based vesicles (see Figure 3.7b) [27].

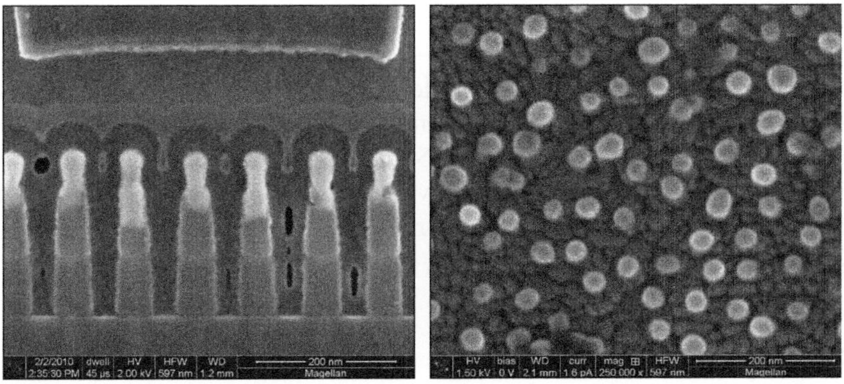

(a) (b)

Figure 3.7 Application results with XHR SEM. (a) Etched cross-section though a flash memory array (Source: FEI). (b) Fullerene-anion based vesicles on indium tin oxide surface. Reprinted with permission from Reference [27]. Homma, *et al.*, *J. Am. Chem. Soc.*, **133** (16), 6364–6370 (2011). Copyright 2011 American Chemical Society. Image courtesy of The University of Tokyo.

Within life sciences there are other challenges with minimizing damage in both sample handling and imaging of often fragile specimens. For example, cryogenic sample preparation and handling can be used to avoid the harsh chemical drying and fixing procedures otherwise required to make biological samples compatible with the SEM. Once in the SEM low dose imaging at low beam energy is critical for the best results on these sensitive samples [28].

3.5 CONCLUSIONS

XHR SEM takes current high-resolution SEM applications and pushes them into a new regime of sub-nm resolution at low beam energies. The beam monochromator makes a significant improvement in SEM capability, without increasing system complexity or compromising ease-of-use or sample size. In addition, it extends the performance of the SEM in the dual-beam FIB/SEM system, as well as in non-immersion lens systems.

For semiconductor manufacturers, this means continuing to use the SEM for more node shrinks than was previous thought possible. For the researcher, it has opened up exciting imaging possibilities for imaging nanoscale structures, providing new and complementary information to that obtained by TEM, AFM [23, 27, 29] and similar techniques.

ACKNOWLEDGEMENTS

The authors are grateful to FEI colleagues for many useful discussions, especially the co-authors of Reference [4], from which this chapter is adapted, and to I. Gestmann, M. Schmidt and K. Murata for images included. We also gratefully acknowledge the support of the Smart Mix Programme of the Netherlands Ministry of Economic Affairs and the Netherlands Ministry of Education, Culture and Science.

REFERENCES

1. L. Reimer, *Image Formation in Low-Voltage Scanning Electron Microscopy*, SPIE Optical Engineering Press, Bellingham, Washington, Volume TT **12** (1993).
2. L. Frank, M. Hovorka, I. Konvalina, *et al.*, Very low energy electron microscopy, *Nucl. Instr. Meth. A*, **645**, 46–54 (2011).
3. R. Young, T. Templeton, L. Roussel, *et al.*, Extreme High Resolution SEM: a paradigm shift, *Microscopy Today*, **16**(4), 24–28 (2008).
4. R. Young, S. Henstra, J. Chmelik, *et al.*, XHR SEM: enabling extreme high resolution scanning electron microscopy, *Proc. SPIE*, **7378**, 737803 (2009).
5. L.Y. Roussel, D.J. Stokes, I. Gestmann, *et al.*, Extreme high resolution scanning electron microscopy (XHR SEM) and beyond, *Proc. SPIE*, **7378**, 73780W (2009).
6. J.R. Michael, D.C. Joy and B.J. Griffin, Challenges in Achieving High Resolution at Low Voltages in the SEM, *Microsc. Microanal.*, **15** (Suppl. 02), 660–661 (2009).
7. J.E. Barth and P. Kruit, Addition of different contributions to the charged particle probe size, *Optik* **101**, 101–109 (1996).
8. O. Scherzer, Sphärische und chromatische Korrektur von Elektronenlinsen, *Optik*, **2**, 114–132 (1947).
9. J. Zach and M. Haider, Aberration correction in a low voltage SEM by a multipole corrector, *Nucl. Instr. Meth. A*, **363**, 316–325 (1995).
10. D. Maas, E. Bosch, J. Horikx, *et al.*, Electrostatic Aberration Correction in Low-Voltage SEM, *Proc. CPO-07*, 84–85 (2006).
11. K. Hirose, T. Nakano and T. Kawasaki, Automatic aberration-correction system for scanning electron microscopy, *Microelectron. Engin.*, doi:10.1016/j.mee .2010.12.040 (2011).
12. H. Kazumori, K. Honda, M. Matsuya, *et al.*, Field Emission SEM with a Spherical and Chromatic Aberration Corrector, *Microsc. Microanal.*, **10** (Suppl. 02), 1370–1371 (2004).
13. S. Uno, K. Honda, N. Nakamura, *et al.*, Aberration correction and its automatic control in scanning electron microscopes, *Optik*, **116**, 438–448 (2005).
14. L.W. Swanson and G.A. Schwind, Review of ZrO/W Schottky Cathode, in *Handbook of Charged Particle Optics, 2nd edn*, CRC Press, Boca Rotan, FL, (2009).

15. H.W. Mook and P. Kruit, Construction and characterization of the fringe field monochromator for a field emission gun, *Ultramicroscopy*, **81**, 129–139 (2000).
16. ISO/24597 measurement methods for SEM image sharpness. Derivative Method.
17. B. Rieger and G. van Veen, Method to determine image sharpness and resolution in scanning electron microscopy images, *Proc. EMC '08*, **1**, 613–614 (2008).
18. E. Kieft and E. Bosch, Refinement of Monte Carlo simulations of electron–specimen interaction in low-voltage SEM, *J. Phys. D: Appl. Phys.*, **41**, 215310 (2008).
19. I. Müllerová and L. Frank, Very Low Energy Scanning Electron Microscopy, in *Modern Research and Educational Topics in Microscopy*, FORMATEX, Badajoz, Spain, 795–804 (2007).
20. H.J. Fitting, E. Schreiber, J.-Ch. Kuhr and A. von Czarnowski, Attenuation and escape depths of low-energy electron emission, *J. Electron Spectrosc. Relat. Phenom.*, **119**, 35–47 (2001).
21. R.M. Tromp, J.B. Hannon, A.W. Ellis, *et al.*, A new aberration-corrected, energy-filtered LEEM/PEEM instrument. I. Principles and design, *Ultramicroscopy*, **110** (7), 852–861 (2010).
22. R. Young, R. Lehmann and I. Gestmann, XHR SEM Bolsters Photolithography Process Control, *Semiconductor International*, April 2010, 14–16 (2010).
23. T. Hachisu, W. Sato, S. Ishizuka, *et al.*, Injection of synthesized FePt nanoparticles in hole-patterns for bit patterned media, *J. Magn. Magn. Mater*, doi: 10.1016/j.jmmm.2010.12.023 (2011).
24. I. Carabante, M. Grahn, A. Holmgren and J. Hedlund, In situ ATR-FTIR studies on the competitive adsorption of arsenate and phosphate on ferrihydrite, *J. Colloid Interface Sci.*, **351** (2), 523–531 (2010).
25. E. Potapova, M. Grahn, A. Holmgren and J. Hedlund, The effect of calcium ions and sodium silicate on the adsorption of a model anionic flotation collector on magnetite studied by ATR-FTIR spectroscopy, *J. Colloid Interface Sci.*, **345** (1) 96–102 (2010).
26. Z. Yermekova, Z. Mansurov and A. Mukasyan, Influence of precursor morphology on the microstructure of silicon carbide nanopowder produced by combustion syntheses, *Ceram. Int.*, **36** (8), 2297–2305 (2010).
27. T. Homma, K. Harano, H. Isobe and E. Nakamura, Preparation and Properties of Vesicles Made of Nonpolar/Polar/Nonpolar Fullerene Amphiphiles, *J. Am. Chem. Soc.*, **133** (16), 6364–6370 (2011).
28. B. Lich, Expediting the Discovery Process with an SEM Work Flow Solution for Life Sciences, *American Laboratory*, April 08 (2011).
29. D. Greif, D. Wesner, J. Regtmeier and D. Anselmetti, High resolution imaging of surface patterns of single bacterial cells, *Ultramicroscopy*, **110** (10), 1290–1296 (2010).

4

The Application of Low-Voltage SEM—From Nanotechnology to Biological Research

Natasha Erdman[1] and David C. Bell[2]
[1]*JEOL USA Inc., USA*
[2]*School of Engineering and Applied Sciences, Harvard University, USA*

4.1 INTRODUCTION

The idea behind nanotechnology is simple: exploit the properties of the nanoscale for the macroscopic world in which we live. Historically, nanomaterials have been with us a long time we just didn't know them as such. The discovery of various nanomaterials was only made possible by electron microscope observation. Although the attribution of who was first remains a debated issue, Sumio Iijima's report in 1991 (Iijima, 1991) is important because it brought carbon nanostructures to the awareness of the whole community. The ability to image carbon, a low atomic number material, is after all an impressive feat, which has led to the imaging of fullerenes and most importantly graphene, which has a potential to become "the new silicon" for future electronics devices. Microscopy has played an integral role in characterization of nanomaterials, and often provides invaluable insight into the structure-properties relationship.

Low Voltage Electron Microscopy: Principles and Applications, First Edition.
Edited by David C. Bell and Natasha Erdman.

In the last decade there has been a quantum leap in the ability of the SEM to observe nanomaterials with ultra-high resolution and exceptional surface detail, in particular employing low-voltage SEM. Improvements in objective lens optics with smaller chromatic and spherical aberration coefficients, ability to deal with charging specimens via precise control of the landing energy of impact electrons and electron signal detection through in-column signal filtering or signal collection angle control have opened new avenues for specimen observation. Moreover, improvements in the SEM technology now allow some of the traditionally TEM-based research of biological materials to be conducted in SEM, which greatly facilitates the throughput and opens up new avenues for biomaterials-based research. This chapter will highlight some of these novel applications as well as the detectors and techniques used to observe nanoscale structures.

4.2 SPECIMEN PREPARATION CONSIDERATIONS

No discussion about application of the SEM for any type of material investigation would be complete without addressing the issues associated with sample preparation. A recent text by Echlin (2009) provides a comprehensive overview of various specimen preparation techniques for SEM and microanalysis. The majority of the samples that end up in the SEM are not ideal; often times they can be non-conductive and sensitive to beam irradiation (for a more detailed discussion of radiation damage mechanisms, see for example; Isaacson, 1975; Reimer and Schmidt, 1984; Talmon, 1987). The use of low voltage SEM can help alleviate some of these issues, as discussed in Chapter 2. However, for some specimens the charge balance is very difficult to achieve even by using very low voltages, due to several factors:

1. *Specimen mounting*—oftentimes users assume that placing a specimen onto a stub with adhesive carbon tape provides secure mounting and conductive pathway. However, this is often not the case, and even supposedly conductive specimens can appear as drifting and charging under the electron beam if mounted in this fashion. It is imperative that a user provides a necessary conductive pathway that connects the specimen with the SEM holder, often times using conductive paint (carbon or silver). The caveat of using carbon tape is that the tape adhesive tends to take a while to outgas and "settle" under vacuum; therefore, high-resolution imaging may not be attainable for

a long time due to specimen drift and charging. The conductivity of carbon tape is also not ideal and depends on the manufacturer, and although no studies exist, it is thought that the conductivity decreases at lower voltage.

2. *Large variations in topography* (inhomogeneous charge distribution). In this case, application of conductive coating may be required, especially if further EDS analysis (higher kV) is needed.
3. *Specimen size*—specimens come in all shapes and sizes, and even though the user's instinct is to look at the specimen as a whole, often times a smaller specimen will satisfy charge balance criterion whereas a larger specimen will cause issues. A great example of this is powder specimens; it is recommended that such specimens should be well dispersed on an SEM specimen stub prior to imaging. This can be accomplished by either dispersing the powder in a solvent and putting a drop of the solution on a stub, or by sprinkling a small amount of powder onto a stub and removing all excess material by using compressed air.

Application of a very thin conductive coating can allow more stable SEM imaging as well as observation at slightly higher kVs for X-ray analysis. The methods for applying conductive coating via either sputtering or evaporation have been available for many years (Goldstein *et al.*, 2003), with choices of metal (gold, palladium, platinum, iridium, chrome, etc.) or carbon. For high-resolution work the metals of choice are typically iridium or chromium due to small grain size (though one has to be careful with respect to specimen storage as the coating will oxidize over time), or evaporated thin layer of carbon which can be somewhat advantageous since it doesn't introduce any obvious artifacts for either imaging or microanalysis.

Ion beam based sample preparation methods may be advantageous when a specimen cross-section is of interest. Focused ion beam (Giannuzzi and Stevie, 2005) and broad argon ion beam (Erdman *et al.*, 2006; Hauffe, 1991) instruments can be particularly useful for preparation of thin film cross-sections, sectioning of small particles, investigation of nanocrystalline interfaces as well as preparation of thin lamellas for STEM-in-SEM.

Preparation of biological, organic and polymer samples typically requires infusion of solutions or vapors containing heavy metal salts, such as osmium or ruthenium tetroxide, uranyl acetate or phosphotungstic acid (Goldstein *et al.*, 2003; Schatten and Pawley, 2008). This preparation can be very time consuming and can introduce artifacts, though it is often essential for its ability to provide contrast enhancement of various features and to allow imaging without need for any additional conductive coating.

4.3 NANOMATERIALS APPLICATIONS

4.3.1 Nanoparticles, Nanotubes and Nanowires

Gold nanoparticles have long been known to be the component on medieval colored glass with different sizes showing as different hues of red or gold, but wasn't until the advent of powerful enough microscopes that the reason for color was known. Currently nanosized gold particles are used for anything from immunolabeling to catalysis. Figure 4.1 shows an example of Au nanosized powder imaged at 0.5 kV using a low-angle backscatter detector (LABe). The powder specimen was prepared by milling the surface with Ar^+ ion beam using a Cross-section Polisher (Erdman *et al.*, 2006). Such preparation assures strain-free specimen cross-section, thus allowing observation of particle crystallography, including twin planes, even at low-voltages with few nm resolution. When the sample is positioned at short working distance, the LABe

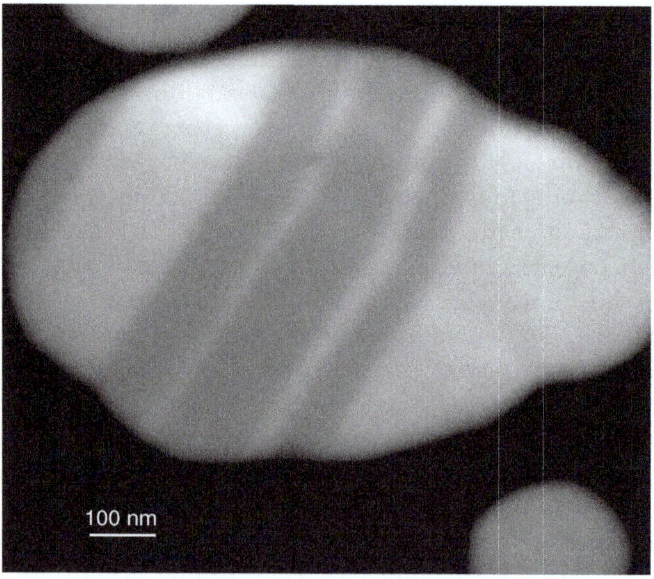

Figure 4.1 Image of Au nanoparticle cross-section taken at 0.5 kV with LABe detector.

detector collects strictly the low-angle backscatter electrons (BSE), which convey high spatial resolution information. Although low-angle BSEs are generated throughout the entire interaction volume of the specimen, they can escape only from the top few nm, thus producing a surface-specific backscatter image. By lowering the accelerating voltage down to few kV or even few hundred Volts, the beam/specimen interaction volume is significantly reduced, yielding images with unprecedented crystallographic surface detail.

Another group of specimens often times examined with FE-SEMs are sub-micron size particles (often oxide-based) that are decorated with heavier element (typically Pt, Pd or Au) nanoclusters serving as a potential candidate for catalysis and other applications. This type of specimen requires utilization of a variety of detectors available on FE-SEM to observe either specimen morphology (SE signal) or distribution of heavier Z component (BE signal) or some combination of both at high special resolution, since the particles are often times a few nm in diameter. The substrate nanopowder also adds an aspect of complexity, since oxides are typically highly charging in SEM; therefore, low kV observation is a must. The user can also utilize FE-SEM's in-column signal filtering capability to distinguish between the nanoclusters and substrates and investigate high Z particle distribution. For example, an energy selective r-filter (JEOL) can be combined with an in-lens SE detector and is capable of continuously adjusting the bias voltage of the accelerating/retarding electrodes placed inside the objective lens to select the energy band of the electrons collected by the in-lens detector (see Chapter 2). The r-filter can thus provide morphological (SE), compositional (BSE) or a fractional blend of the two signals. An example of Au nanoclusters on TiO_2 powder sample is shown in Figure 4.2. The clusters were imaged at 2 kV using several different filter options: (a) Sb setting (suppressing very low energy SEs that contribute to charging and collecting SE signal with small fraction of higher energy electrons), that shows both morphology of the individual TiO_2 particle surface and the Au clusters covering the surface; (b) Bs setting, which begins to suppress low energy contribution in favor of additional high energy signal to show more pronounced contrast from the Au clusters; and (c) BE setting, giving essentially only compositional information. The in-lens filter clearly resolves 2–5 nm individual Au clusters on the oxide surface and allows observation of their overall distribution across the particle surface without charging effects at 2 kV. Moreover, the BE image clearly shows that the large cluster (identified with an arrow) is

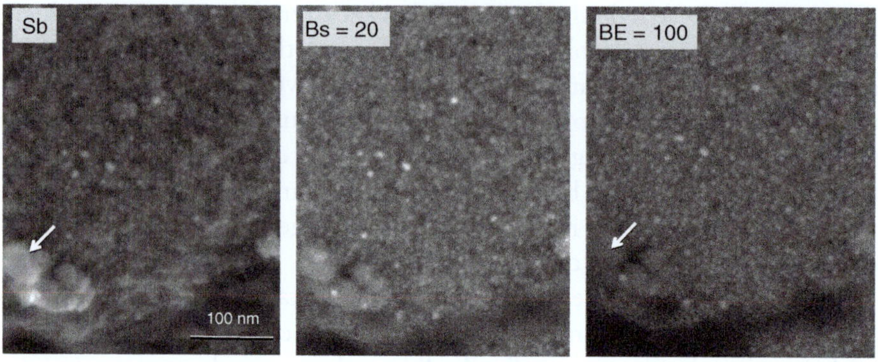

Figure 4.2 Au nanoparticles and clusters on TiO_2 imaged with r-filter on JSM-7600F.

in fact part of the support TiO_2 matrix since it does not exhibit strong contrast like the rest of the Au clusters.

Synthesis of material with a desired crystal structure is a major challenge in materials engineering. Investigation of nanowire nucleation and growth and the crystallographic relationship between nanowires and substrates can also be performed using low voltage FE-SEM. For instance, InAs as a semiconductor with high electron mobility, especially in nanostructures such as quantum dots and nanowires, is being explored for applications in optoelectronic and high-speed electronic devices (Sarukura *et al.*, 1998; Pettersson *et al.*, 2006). A sample of InAs nanowires (wurtzite structure) synthesized via chemical beam epitaxy on InAs zinc blend substrate (Bao *et al.*, 2009) was investigated in FE-SEM. For comparison the wires were imaged using both in-lens and below the lens SE detectors at high (15 kV) and low (1 kV) accelerating voltage (see Figure 4.3). The images show that InAs nanowires grow vertically and at an angle from the substrate. Due to the conditions for the formation of the nanowires, the wurtzite crystal structure is preferred, in contrast to the zinc-blende crystal structure that is energetically favorable for bulk InAs (Ohlsson *et al.*, 2002; Mandl *et al.*, 2006). A pyramid-like base forms at the root of each InAs nanowire during the fabrication process. The images clearly demonstrate that low kV images provide significantly enhanced surface detail as compared to high kV images, where the surface features are indistinguishable due the overwhelming signal collected from the large interaction volume and areas deep within the InAs substrate. The low kV in-chamber secondary detector (LEI) images clearly show the more traditional SE image with distinct topographic contribution to the image, resulting from the overall

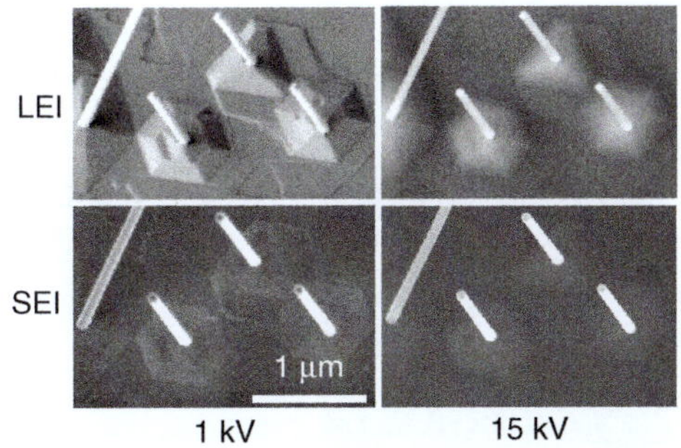

Figure 4.3 Comparison of in-chamber (LEI) and in-lens (SEI) detector performance at high and low kV for a sample of InAs nanowires.

geometry of the sample with respect to the detector position in the SEM chamber. The in-chamber detector is more sensitive to the so-called SE_{II} electrons that are generated by multiple cascades deeper inside the interaction volume and, therefore, have higher energy. The in-lens SEI detector geometry, on the other hand, offers top-down view of the sample with minimal contribution from specimen topography and can separate SE_I/SE_{II} electrons vs. higher energy/high angle backscatter electrons using energy filter (r-filter). The nanowires show strong contrast as compared to substrate due to their orientation, since the SE yield is dependent on the specimen tilt via secant law $\delta(\theta) = \delta_0 \sec(\theta)$. When a specimen has a cylindrical shape (such as a nanowire) and lying perpendicular to the incident beam, the number of secondary electrons SE_I generated from the top surface is significantly smaller than from the side surface SE_{II}, based on the beam penetration geometry and additional SE_{II} generation due to backscattered electrons, which are exiting the side surface (see Figure 4.4). Therefore, the wires that are lying down flat on the sample surface as well as the substrate surface itself will have lower contrast in the image and the wires that are oriented standing almost perpendicular with respect to the sample surface will have higher contrast. Furthermore, the in-lens detector employed at short working distances (see Figure 4.4) allows high-resolution imaging of the wire surface as well as the Au catalyst seed particle on top of the nanowire. This image was collected at 0.5 kV to further enhance surface detail.

Nanoparticles, nanowires and nanotubes can also be easily observed in the FE-SEMs utilizing a STEM-in-SEM technique (see Chapter 2). In this configuration, the specimen is mounted on a TEM grid with carbon support film and is imaged in transmission by utilizing either a dedicated STEM detector or a so-called STEM converter. This technique

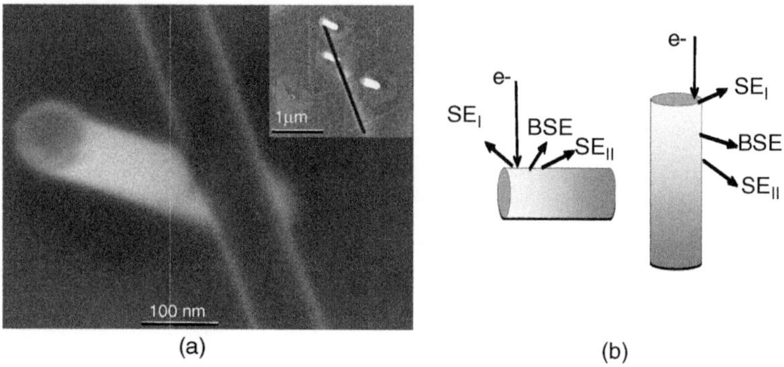

Figure 4.4 (a) Image of the InAs nanowires collected at 0.5 kV with in-lens SE detector. The inset shows lower magnification image with enhanced surface detail. (b) Image formation mechanism for flat versus standing nanowire.

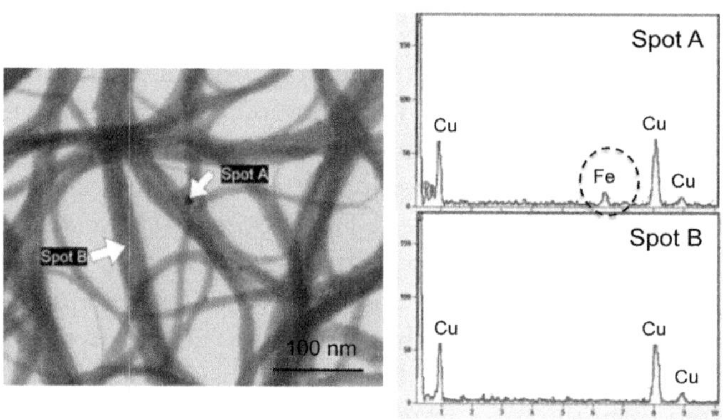

Figure 4.5 Left: image of carbon nanotubes in STEM mode taken with JSM-7600F at 30 kV. Right: spot EDS analyses of two different locations: catalyst particle (Spot A), and CNT matrix (Spot B) (See plate section for coloured version).

provides effectively a low kV bright filed or dark-field STEM image (typically at 30 kV), resulting in substantially reduced knock-on damage as compared to traditional high-kV TEM or STEM imaging. Figure 4.5 shows an example of carbon nanotubes (CNTs) imaged in the STEM mode, with the Fe catalyst particles clearly observed. An advantage to imaging a specimen like this in STEM mode is the ability to perform EDS spot analysis (Figure 4.5) to characterize the individual 5–10 nm catalyst particles, which is nearly impossible with a bulk specimen even at low kV. The EDS spectrum clearly identifies a catalyst particle (Fe) on the CNT.

4.3.2 Nanoporous Materials

Nanoporous materials such as zeolites and mesoporous crystals have attracted a lot of attention in recent years. These types of materials have been widely used in the field of catalysis (Bhatia, 1990; Chen *et al.*, 1996; Wietkamp, 2000), and are gaining prominence for researchers' ability to alter their properties via incorporation of various metal nanoparticles and organic molecules within the pore network. In order to precisely control this new functionality of nanoporous materials it is crucial to understand their structure, growth mechanisms and composition. Traditionally, nanoporous materials have been examined either in TEM for structure determination via diffraction and HRTEM (Dorset, 2006; Rice, 1990; Treacy *et al.*, 1986; Wagner *et al.*, 1999) or in SEM at relatively low magnification for overall particle morphology. The recent advances in low kV SEM technology have been instrumental in the ability to observe the fine surface structure of these pore networks (Che *et al.*, 2003; Stevens *et al.*, 2008; Stevens *et al.*, 2009; Terasaki *et al.*, 2004; Tuysuz *et al.*, 2008; Wakihara *et al.*, 2005), twinning and growth features in zeolites (Terasaki, 1999), and other features, especially since majority of these materials are insulating. Moreover, the ability to filter collected signals and high-resolution low kV BSE imaging allow observation of compositional differences as well as precise location of nanoparticles within the individual pores.

The charging component is one of the main contributors to the loss of information when observing nanoporous materials. The effect of charging manifests itself as either "flattening" of the image due to the beam deflection close to the source of charging (in this case a nanoporous particle), or extremely high or low contrast and image distortion. For example, when observing mesoporous silica (SBA-15) crystals, even at 0.5 kV the surface detail is completely obscured in the resulting image (see Figure 4.6a). Since the charging is a result of a build-up of an inhomogeneous electric field generated when source electrons impact the specimen surface, it may be possible to balance the charge via precise control over the ratio of emitted to impact electrons (see Chapter 2) using specimen bias. When specimen bias is applied, the surface structure can be easily observed at the same "landing" voltage of 0.5 kV (Figure 4.6b); in this case the gun voltage is 2 kV, the specimen bias is 1.5 kV. The HRSEM images show the surface with nanoporous channels that terminate in either U-shaped (Figure 4.6c) or open (Figure 4.6d) configuration. Using specimen bias for observation of

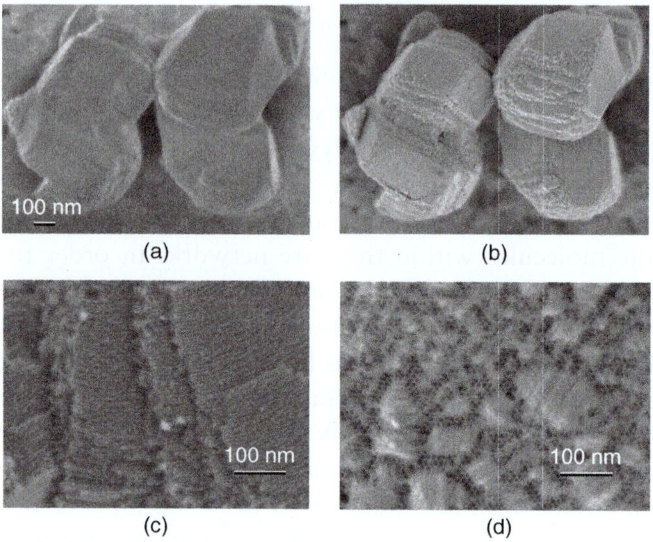

Figure 4.6 Specimen of SBA-15 imaged at 0.5 kV with JSM-7600F. (a) no specimen bias applied; (b) specimen bias of 1.5 kV; (c) U-shape channel surface termination; (d) open channel surface termination.

nanoporous materials provides charge balancing while maintaining high resolution and surface sensitive imaging at low accelerating voltages.

Another avenue for imaging nanoporous materials is by using the STEM-in-SEM technique. Though the resolution of the STEM-in-SEM (around 0.6 nm) is about five times lower than the conventional TEM, it is possible with the current state of the art FE-SEMs to observe sub-nanometer pore structures in bright-field (Figure 4.7a), as well

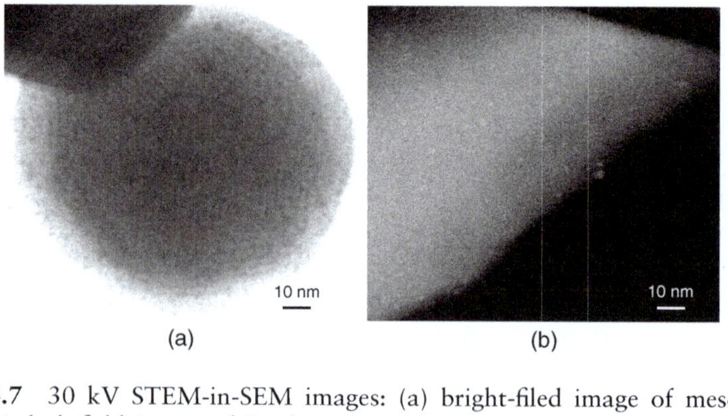

Figure 4.7 30 kV STEM-in-SEM images: (a) bright-filed image of mesoporous silica; (b) dark-field image of Pt clusters on alumina support (Courtesy of JEOL USA Inc).

as nanoparticle distribution on the catalyst supports in dark-field (Figure 4.7b). The specimen preparation becomes a key component in this case, since the specimens have to be very thin for adequate beam penetration at 30 kV, zeolite powders will typically have to be prepared using a microtome.

4.3.3 Graphene

Graphene is the hexagonal arrangement of carbon atoms forming a one-atom thick planar sheet and has outstanding properties that render it an attractive materials option for electronics applications (Novoselov *et al.*, 2004; Geim and Novoselov, 2007). Traditionally, graphene structures have been imaged either with aberration corrected TEM (see Chapter 7) or using AFM or STM (see for example Lui *et al.*, 2009; Frank *et al.*; 2007). Moreover, research demonstrating the ability not only to image but also perform ion-beam lithography in graphene using He Ion Microscope has been reported (Bell *et al.*, 2009).

Graphene layers, especially if deposited on a solid bulk substrate, can also be imaged and analyzed using low kV FE-SEM. Figure 4.8(a) and

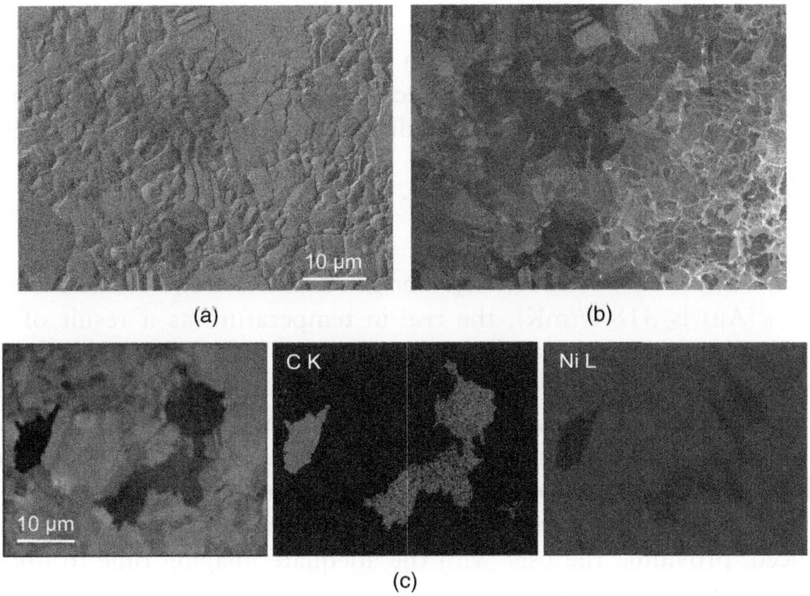

Figure 4.8 Graphene film on Ni substrate. (a) In-chamber SE and (b) in-lens SE images collected at 2 kV (c) in-lens image and EDS maps of the sample collected at 2 kV.

(b) show FE-SEM images taken at 2 kV of a specimen of graphene layers deposited on Ni substrate using two different SE detectors. The in-lens SE image (Figure 4.8b) is the only image actually clearly showing both the graphene layers of varying thickness and the underlying structure of Ni substrate. This is due to the fact that the in-lens SE detector is highly sensitive to surface specific information at low kilovolts, and the graphene layers are limited to few atomic layers on the specimen surface. Figure 4.8(c) shows EDS maps of graphene layers on Ni substrate, obtained at 2 kV, 10 nA of beam current. By performing EDS analysis at 2 kV accelerating voltage, the beam/specimen interaction volume is significantly reduced, so that discrete maps of graphene and Ni components can be resolved. Additional consideration should be given to minimizing analysis time, to prevent any significant buildup of carbon contamination on the sample surface that could obscure characterization; in this case the maps were collected for 2 min. The high beam current setting used for this analysis ensured adequate data collection for compositional interpretation even with a very short acquisition time at low kV.

4.4 BEAM SENSITIVE MATERIALS

Primary electron beam interaction with a specimen and subsequent inelastic scattering processes result in potential heat generation, specimen damage and contamination. The electron beam power (*P*) is distributed within the interaction volume according to the equation (Reimer, 1993):

$$P = -\kappa \cdot 2\pi r^2 \cdot \frac{dT}{dr} \tag{4.1}$$

where κ is thermal conductivity. For good heat conductors like metals (e.g., κ(Au) is 318 W/mK), the rise in temperature as a result of the electron beam impact is insignificant. In the case of materials with low thermal conductance (e.g., κ(polystyrene) is ~0.08 W/mK) the user should expect a more dramatic change in temperature. However, since at low voltages the volume of increased temperature is significantly reduced, lower amount of thermal damage should be expected or at the very least the speed with which the damage occurs should be substantially reduced, providing the user with the adequate imaging time to obtain the necessary information.

Figure 4.9 shows an example of a calcite grain in shale imaged at 2 kV (Figure 4.9a) versus the same grain imaged at 5 kV (Figure 9b–d). The 5 kV images were taken as a sequence of frames, showing accumulated

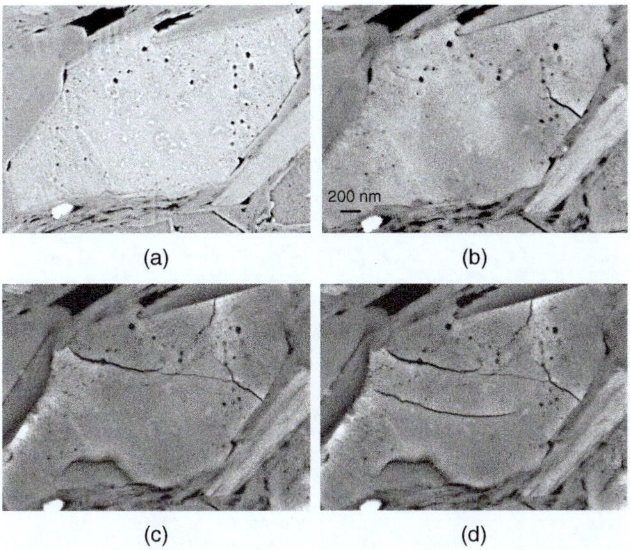

Figure 4.9 Calcite grain in shale imaged at (a) 2 kV versus the same grain imaged at 5 kV (b–d). The 5 kV images (b–d) were taken sequentially to show damage to the crystal developing over beam exposure time. The sample was prepared using JEOL Cross-section polisher.

damage to the grain (such as propagation of cracks and shrinking of the grain) over time with subsequent scans. The damage effectively started within the first few seconds of beam rastering and continued over several minutes until steady state has been reached. Initial imaging the same grain at 2 kV has not shown any obvious damage.

Another example of low voltage imaging of beam sensitive specimen is a sample of PETN (Pentaerythritol tetranitrate). This material is an insulator and often requires conductive coating for imaging or use of ultra-low voltages for observation. Figure 4.10 shows an image of PETN particle cross-section imaged with FE-SEM at 0.5 kV. The image clearly shows the internal network of large pores, as well as smaller nanopores present on the surface. In this case in-chamber E/T secondary detector was used for imaging since it is less sensitive to charging, and frame integration was used for image acquisition to reduce potential specimen charging and beam damage.

4.5 SEMICONDUCTOR MATERIALS

Failure analysis is crucial for semiconductor industry in order to identify potential processing issues or to optimize device performance. Low-k

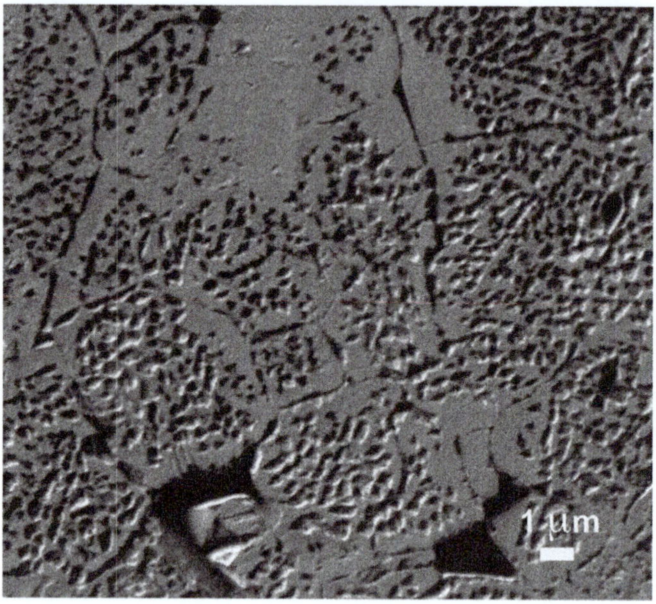

Figure 4.10 PETN particle cross-section imaged with FE-SEM at 0.5 kV (Sample courtesy of Dyno Nobel).

dielectric materials and layers used to reduce the time constant of the chip have always been challenging for SEM imaging, due to their potential charging characteristics and their susceptibility to thermal damage. Figure 4.11 shows the ability of FE-SEM to image these materials with high spatial resolution and minimal damage by using a combination of

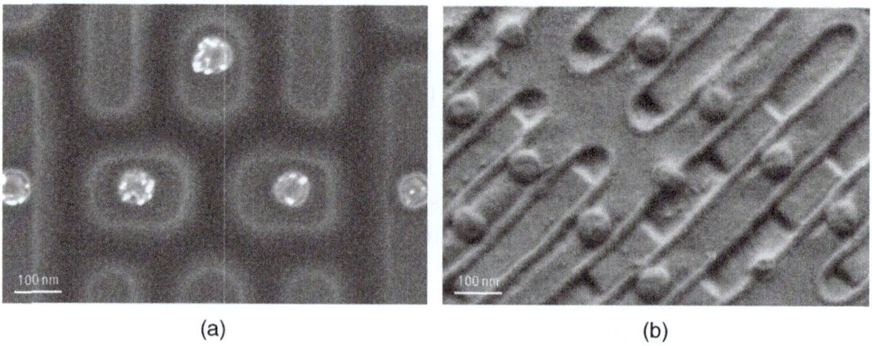

Figure 4.11 (a) Image of a polished semiconductor device at metal 1/via level, 2 kV, 57 pA; (b) Semiconductor device polished to the contact level just before exposing the polysilicon level, 0.3 kV, 70 pA (Images courtesy of Carl Zeiss NTS).

low voltage and reduced beam current. Furthermore, this type of imaging is not limited to flat specimens, the same type of analysis can be performed on specimens at high tilt, allowing further versatility of analysis.

Voltage contrast is an important contrast mechanism for failure analysis imaging of semiconductor devices. In the case of voltage contrast, a portion of semiconductor device that is positively charged (biased) retards low energy SEs and appears darker in the image than the rest of the specimen at ground potential, as compared to the portion of the sample that is negatively charged (biased) and thus repels more SEs that are then collected by SE detector. Use of low accelerating voltage ensures minimal radiation damage to the semiconductor device during observation.

4.6 BIOLOGICAL SPECIMENS

The nanoscale character of many internal cellular structures as well as the need to examine these structures in 3D (via tomography) has over the years guided biologists to use a TEM as their imaging tool. However, the TEM imaging has the major limitation that only a small snapshot of the overall specimen is examined, thus leaving the researchers with the need to examine multiple specimens. An advantage of the SEM for biological research is in its inherent ability to combine both macro- and nanoscale observation, providing the researcher with the overview of the morphology of the biological structure of interest (for example an individual cell) using SE mode, and if necessary combine this with the BSE compositional image of the immunocytochemical markers or stains employed to highlight a particular nanoscale aspect of the structure. The current resolution of FE-SEMs allows for imaging of immuno-gold markers down to 2 nm with SE/BSE detectors and down to sub-nm scale with STEM-in-SEM technique. Moreover, the advantage of low voltage, as has been pointed in Chapter 1, is the ability to image uncoated biological specimens without the need to employ techniques like low vacuum or Environmental SEM (see for example Stokes, 2008). Low-voltage obtains extremely surface sensitive images at high spatial resolution, with minimal charging. Figure 4.12 shows an example of a pollen spore imaged at 100 V without any need for conductive coating. In this case the charge balance was achieved by employing 2 kV specimen bias (Gentle Beam). Note preservation of large depth of focus and fine surface detail obtained with these imaging conditions.

Comparative imaging of sectioned biological materials with secondary, backscattered and transmitted electrons in the SEM has been

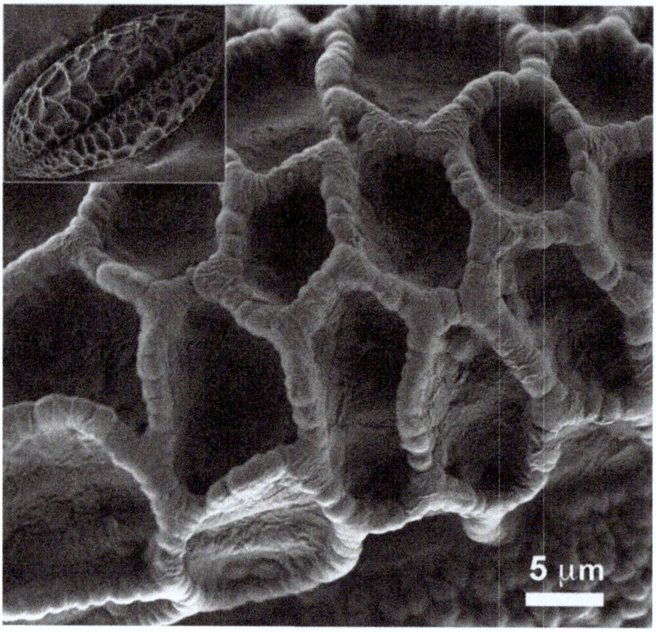

Figure 4.12 SE image of uncoated pollen specimen taken at 100 V. Inset: low magnification image of the same spore.

available to investigators since the mid-1970s (Miyai *et al.*, 1976). Backscatter electron and scanning transmission electron imaging in conventional SEM yielded information that was promising; however, little was done to improve neither preparative procedures nor the utilization of more sensitive detection systems since the resolution of an SEM at that time did not approach that of the TEM. With the development of field emission scanning electron microscopes in the late 1980s a resurgence of interest in the backscatter imaging of biological samples occurred with the development of immuno-gold surface techniques (de Harven *et al.*, 1984). This was limited to surface epitope labeling, and with the advent of the confocal microscope backscatter imaging in the FE-SEM again became an underutilized tool in biological imaging.

Interest in backscatter imaging of sectioned biological material is currently being revisited with the development of higher resolution backscatter detectors, in-lens detectors and energy filters in the FE-SEM. Of particular interest are the low voltage techniques for surface sensitive analysis of nanostructures (for example, imaging bacteriophages or the tip links connecting stereocilia and receptor cells in the inner ear), as well as utilization of low voltage SEM and STEM-in-SEM to examine ultra-thin sections of tissue in either two- or 3D.

One of the particularly interesting avenues of research is utilizing the technological developments in FE-SEM electron optics and detector system to image very large areas of thick sections (Ackerley, 2005), serial microtomed surfaces of tissue blocks (Denk and Horstmann, 2004) and serial sections of tissue using novel ultramicrotomy techniques (Kasthuri *et al.*, 2007) for either correlative microscopy (combining SEM with fluorescence or confocal data) or serial reconstruction in 3D. The large FE-SEM specimen chamber is capable of accommodating these large specimens and with significantly improved SEM electron optics images can easily be acquired at near TEM resolutions. The low voltage operation (1.5–3 kV) combined with imaging using an in-lens detector with filtering grid results in highly surface sensitive as well as compositionally sensitive images. An example shown in Figure 4.13 is an image of a 50 nm brain stem section, clearly showing not only the myelin sheath layers but also the immuno-gold labels (5 nm) at the resolution previously attributed only to TEM imaging. The image was taken with

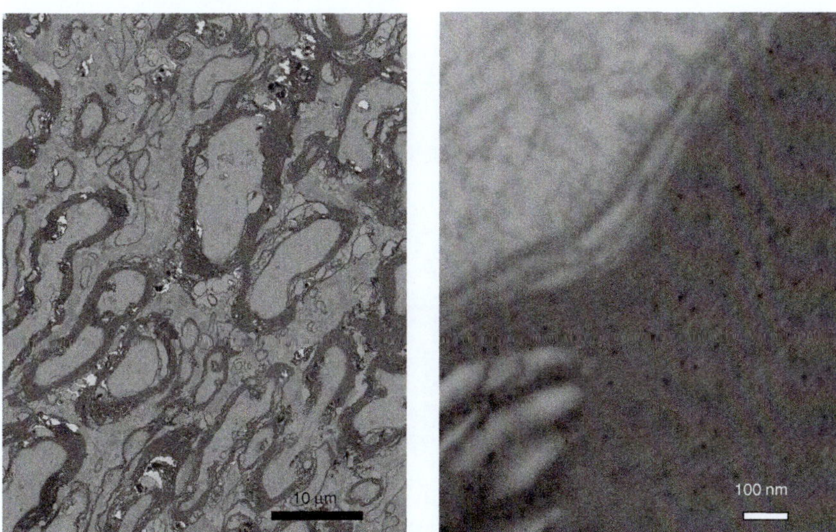

(a) (b)

Figure 4.13 (a) Low magnification image of the brain stem thin section taken at 2 kV using through the lens detector on JSM-7800F; (b) high resolution image of the sample showing myelin layers and immuno-gold labels. The contrast for both images was inverted. Images courtesy of Mr M. Shibata (JEOL USA).

an in-lens detector with energy filter set up to collect only backscatter signal, and the contrast inverted to produce a more familiar contrast of a typical TEM image of this type of specimen.

In addition, STEM detectors can easily be interfaced with FE-SEM and ultrathin sections mounted on grids can be observed using the same optics as regular SEM operation. The advantage of STEM-in-SEM for biological materials is that at low voltages (30 kV and lower), the biological materials that previously appeared featureless when using an 80–120 kV TEM (unless enhanced by fixation, staining and labeling with heavy metal salts), can now be revealed by utilizing the physics of mass-thickness contrast with minimal preparation procedures.

An example of this is given in Figure 4.14: the image shows a negatively stained (1% Uranyl Acetate) T4 bacteriophage, approximately 200 nm long and 90 nm wide. The image clearly shows not only the overall structure of the phage, but also the tail fibers that guide the binding of the phage to the cellular surface of the bacteria. Previously such images could only be acquired via TEM.

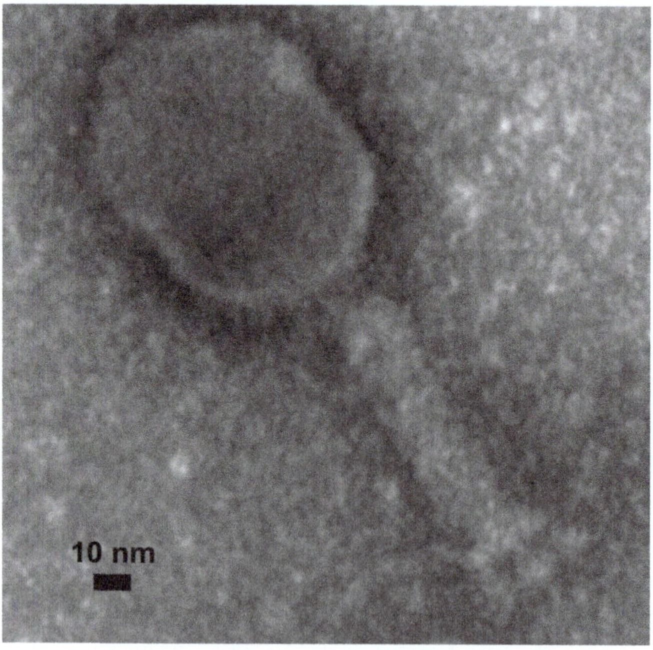

Figure 4.14 Image of negatively stained T4 bacteriophage imaged with 30 kV STEM-in SEM.

The examples highlighted above are just a few avenues of research into biological structures using the improved electron optics and detectors of the new FE-SEMs operated in low voltage regime. For further reading, a comprehensive review of the biological research in low-voltage SEM is presented in the recent volume by Schatten and Pawley (2008), with main focus on cell biology and sample preparation protocols, cryogenic techniques and correlative microscopy.

4.7 LOW-VOLTAGE MICROANALYSIS

The ability of an FE-SEM to increase probe current for fast microanalysis (minutes vs hours or days), while still maintaining a small probe size and small volume of excitation for high spatial resolution has been the holy grail of microanalysis in SEM. This is particularly true for low-voltage microanalysis of nanostructured materials. Traditionally, microanalysis in SEM via either EDS or WDS has been limited to few hundreds of nm spatial resolution at best; however, new classes of nanomaterials require an order of magnitude better data.

To address these challenges, current electron microscope designs have the ability to employ additional lenses to help produce smallest spot sizes with corresponding large currents. For example, the Aperture Angle Control Lens produces a 3 nm probe at 5 kV with 5 nA beam current, more than sufficient current/spot size combination for quick and high-resolution microanalysis (additional details on this technology can be found in Chapter 2). Figure 4.15 shows EDS maps of Ni pillars grown on ITO (indium tin oxide) substrate; the map was collected at 3 kV for 5 min using a 30 mm^2 SDD EDS detector. Each of the pillars measures approximately 50–60 nm in diameter, and they are clearly resolved in the Ni map; moreover, the spaces between the pillars

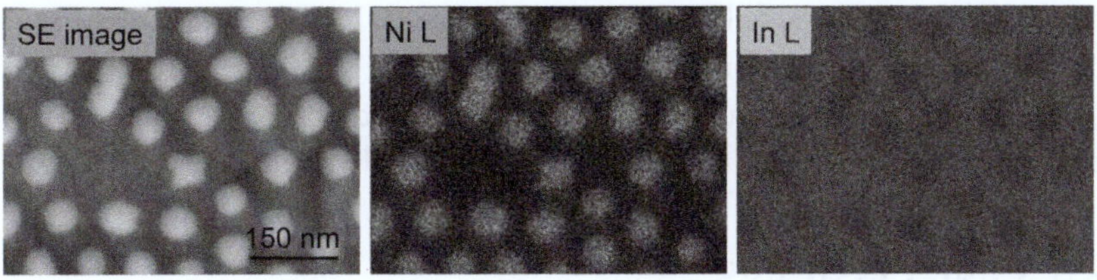

Figure 4.15 Ni pillars on ITO (indium tin oxide). 3 kV EDS mapping (30 mm^2 Oxford SDD), 5 min acquisition with drift correction (See plate section for coloured version).

Figure 4.16 Slip planes in diamond imaged with Centaurus KE detector at 2 kV. The sample was prepared using JEOL Cross-section Polisher.

(~50 nm) correlating to the ITO substrate indium signal are clearly resolved as well.

Another advantage of the combination of small spot size with large probe current is the opportunity to revisit analytical techniques that were previously reserved for high accelerating voltages. One such technique is cathodoluminescence (CL), where the electron beam impacting the specimen surface induces electron hole pair recombination resulting in emission of photons with a characteristic spectral signature depending on the material and defect structures. Use of a simple panchromatic CL detector (the photons are collected by an elliptical mirror and detected by a photomultiplier tube detector) allows imaging of defect structure that would be difficult to observe by using strictly SE or BSE emission. An example of such analysis is shown in Figure 4.16, where the use of CL provides observation of slip planes in diamond. In this case, use of higher accelerating voltages would blur out the surface signal and while providing the information about cathodoluminescence signature of diamond grains, would not show individual features with adequate sharpness.

4.8 CONCLUSIONS

Microscopy has played an integral role in characterization of nanomaterials, one might say that electron microscopy was and is the leading technique for nanomaterial analysis. It is a lot easier to understand something if you can see it and can provide invaluable insight into the structure-properties relationship of the material.

The ability of the SEM to observe nanomaterials with ultra-high resolution and exceptional surface detail, in particular low-voltage SEM have been due in part to improvements in objective lens optics and the ability to deal with charging specimens via precise control of the landing energy of primary electrons and improvements in electron signal detection.

Advances in instrumentation continue to push the boundaries of what is possible with electron microscopy and the resolution and analytical capabilities of SEM continue to improve in the low-voltage regime.

REFERENCES

C.A. Ackerley, A. Tilups and C. Nielsen, "Wien–type" filtered secondary electron imaging of immunocytochemically labeled methacrylate sections of tissues in a field emission scanning electron microscope (FESEM), *Microsc. Microanal.*, **11**, 1096 (2005).

J. Bao, D.C. Bell, F. Capasso, *et al.*, Nanowire induced thin-film Wurtzite InAs structures on zinc-blende InAs substrate, *Adv. Mat.*, **21**, 3654–3658 (2009).

D.C. Bell, M.C. Lemme, L.A. Stern, *et al.*, Precision cutting and patterning of graphene with helium ions, *Nanotechnology*, **20**, 455301 (2009).

S. Bhatia, *Zeolite catalysis: principles and applications*, CRC Press, Boca Raton, FL (1990).

S. Che, K. Lund, T. Tatsumi, *et al.*, Direct observation of 3D mesoporous structure by scanning electron microscopy (SEM): SBA-15 silica and CMK-5 carbon. *Ang. Chemie*, **42**, 2182–2185 (2003).

N.Y. Chen, W.E. Garwood, and F.G. Dwyer, *Shape selective catalysis in industrial applications*, Marcel Dekker, NY (1996).

E. de Harven, R. Leung, and H. Christensen, A novel approach for scanning electron microscopy of colloidal gold-labeled cell surfaces, *J. Cell Biol.*, **99**, 53 (1984).

W. Denk and H. Horstmann, Serial block-face scanning electron microscopy to reconstruct three-dimensional tissue nanostructure, *Plos. Biology*, **2**, 1900 (2004).

D.L. Dorset, The role of electron diffraction in zeolite structure determination, *Acta Cryst.*, **A62**, s47 (2006).

P. Echlin, *Handbook of Sample Preparation for Scanning Electron Microscopy and X-Ray Microanalysis*, Springer (2009).

N. Erdman, R. Campbell, and S. Asahina, Precise SEM cross-section polishing via argon beam milling, *Microsc. Today*, **14**, 22–25 (2006).

I.W. Frank, D.M. Tanenbaum, A.M. van der Zande, and P.L. McEuen, Mechanical properties of suspended graphene sheets, *J. Vac. Sci. Technol. B*, **25**, 2558 (2007).

A.K. Geim and K.S. Novoselov, The rise of graphene, *Nature Materials*, **6**, 183–191 (2007).

L.A. Giannuzzi and F.A. Stevie (eds), *Introduction to Focused Ion Beams: Instrumentation, Theory, Techniques, and Practice*, Springer, New York (2005).

J. Goldstein, D.E. Newbury, D.C. Joy, *et al.*, *Scanning electron microscopy and x-ray microanalysis*, 3rd edn, Springer, New York, (2003).

W. Hauffe, Production of microstructures by ion beam sputtering, in *Sputtering by particle bombardment III. Topics in Applied Physics*, **64**, Springer, Berlin (1991).

S. Iijima, Helical microtubules of graphitic carbon, *Nature*, **354**, 56–58 (1991).

M. Isaacson, Inelastic scattering and beam damage of biological molecules, in *Physical Aspects of Electron Microscopy and Microbeam Analysis*, B. Siegel (ed.), 247–258 (1975).

N. Kasthuri, K. Hayworth, J. Lichtman, *et al.*, New technique for ultra-thin serial brain section imaging using scanning electron microscopy, *Microsc. Microanal.*, **13**, 26 (2007).

C.H. Lui, L. Liu, K.F. Mak, *et al.*, Ultraflat graphene, *Nature*, **462**, 339–341 (2009).

B. Mandl, J. Stangl, T. Mårtensson, *et al.*, Au-Free Epitaxial Growth of InAs Nanowires, *Nano Lett.*, **6**, 1817–1821 (2006).

K. Miyai, J.L. Abraham, D.S. Linthicum, and R.M. Wagner, Scanning electron microscopy of hepatic ultrastructure: secondary, backscattered, and transmitted electron imaging, *Lab Invest.*, **35**, 369 (1976).

K.S. Novoselov, A.K. Geim, S.V. Morozov, *et al.*, Electric Field Effect in Atomically Thin Carbon Films, *Science*, **306**, 666 (2004).

B.J. Ohlsson, M.T. Björk, A.I. Persson, *et al.*, Growth and characterization of GaAs and InAs nano-whiskers and InAs/GaAs heterostructures, *Phys. E*, **13**, 1126–1130 (2002).

H. Pettersson, J. Trägårdh, A.I. Persson, *et al.*, Infrared Photodetectors in Heterostructure Nanowires, *Nano Lett.*, **6**, 229–232 (2006).

L. Reimer, *Image Formation in Low-Voltage Scanning Electron Microscopy*, SPIE Series, Bellingham, WA (1993).

L. Reimer and A. Schmidt, The shrinkage of bulk polymers by radiation damage in an SEM, *Scanning*, **7**, 47–53 (1984).

S.B. Rice, J.Y. Koo, M.M. Disko, and M.M.J. Treacy, On the imaging of Pt atoms in zeolite frameworks, *Ultramicroscopy*, **34**, 108–118 (1990).

N. Sarukura, H. Ohtake, S. Izumida, and Z. Liu, High average-power THz radiation from femtosecond laser-irradiated InAs in a magnetic field and its elliptical polarization characteristics, *J. Appl. Phys.*, **84**, 654 (1998).

H. Schatten and J.B. Pawley (eds.), *Biological low-voltage scanning electron microscopy*, Springer, New York (2008).

S.M. Stevens, K. Jansson, N.S. John, *et al.*, High-Resolution scanning electron and atomic force microscopies: observation of nanometer features on zeolite surfaces. *Studies in Surf. Sci. and Catal.*, **174** (2), 775–780 (2008).

S.M. Stevens, K. Jansson, C. Xiao, *et al.*, An Appraisal of High Resolution Scanning Electron Microscopy Applied To Porous Materials, *JEOL News*, **44** (1), 17 (2009).

D. Stokes, *Principles and Practice of Variable Pressure/Environmental Scanning Electron Microscopy (VP-ESEM)*, John Wiley & Sons, Ltd, Chichester (2008).

Y. Talmon, Electron beam radiation damage to organic and biological cryo-specimens, in *Cryotechniques in Biological Electron Microscopy*, R.A. Steinbrecht and K. Zerold (eds), Springer-Verlag (1987).

O. Terasaki, T. Ohsuna, Z. Liu, *et al.*, Structural study of meso-porous materials by electron microscopy, *Studies in Surf. Sci. and Catal.*, **148**, 261–288, (2004).

M.M.J. Treacy, J.M. Newsam, R.A. Beyerlein, *et al.*, The Structure of Zeolite CSZ-1 Interpreted as a Rhombohedrally-Distorted Variant of the Faujasite Framework, *J. Chem. Soc. Chem. Commun.*, **15**, 1211–1214 (1986).

H. Tuysuz, C.W. Lehmann, H. Bongard, *et al.*, Direct imaging of surface topology and pore system of ordered mesoporous silica (MCM-41, SBA-15, and KIT-6) and nanocast metal oxides by high resolution scanning electron microscopy. *J. Am. Chem. Soc.*, **130** (34), 11510–11517 (2008).

P. Wagner, O. Terasaki, S. Ritsch, *et al.*, Electron Diffraction Structure Solution of a Nanocrystalline Zeolite at Atomic Resolution, *J. Phys. Chem. B*, **103**, 8245–8250 (1999).

T. Wakihara, Y. Sasaki, H. Kato, *et al.*, Investigation of the surface structure of zeolite A, *Phys. Chem. Chem. Phys.*, 7, 3416–3418 (2005).

J. Wietkamp, Zeolites and catalysis, *Solid State Ionics*, **131** (1–2), 175–188 (2000).

5

Low Voltage High-Resolution Transmission Electron Microscopy

David C. Bell
School of Engineering and Applied Sciences, Harvard University, USA

5.1 INTRODUCTION

Low-Voltage High-Resolution Electron Microscopy (LV-HREM) has several advantages, including increased cross-sections for inelastic and elastic scattering, increased contrast per electron and improved spectroscopy efficiency, decreased delocalization effects and reduced radiation knock-on damage (Bell *et al.*, 2010, 2011; Kaiser *et al.*, 2011; Sasaki *et al.*, 2010). Together, these often improve the contrast to damage ratio obtained on a large class of samples. Third-order aberration correction now allows us to operate the TEM at low energies while retaining atomic resolution, which was previously impossible. At low voltage the major limitation to resolution becomes the chromatic aberration limit.

The history of electron microscope development shows that the trend in electron gun design has been towards brighter emission with less energy spread (Hawkes and Spence, 2007). Thermal emission yields energy spreads on the order of 1.2 eV; the Schottky field emission gun (FEG) yields 0.7 eV, and the cold FEG yields down to the 0.3 eV

Low Voltage Electron Microscopy: Principles and Applications, First Edition.
Edited by David C. Bell and Natasha Erdman.

range. The competing needs for high brightness, low cost and emission stability have meant that cold FEGs are not as commonly available from microscope vendors as thermal or Schottky emission sources. One way to combine the high current and stability of a thermally-assisted field emitter with the low energy spread of a cold FEG is to incorporate a monochromator directly in the source (Rose, 2009). By using a source monochromator (or a chromatic aberration correction) it is, of course, possible to reduce the effect of chromatic aberration and achieve a usable high-resolution limit at low voltage down to less than one Angstrom (Figure 5.1). The combination of objective lens Cs correction and a FEG-source monochromator represents a further step along the path towards an ideal microscope, while expansion of the temporal coherence envelope function by reducing the ΔE is a direct benefit that can lead to improved resolution. Monochromator integration further benefits applications that specifically depend on energy resolution, such as spectroscopy and "atomic-scale" energy-filtered imaging. By reducing the primary beam voltage of the TEM, there are distinct gains in using electron

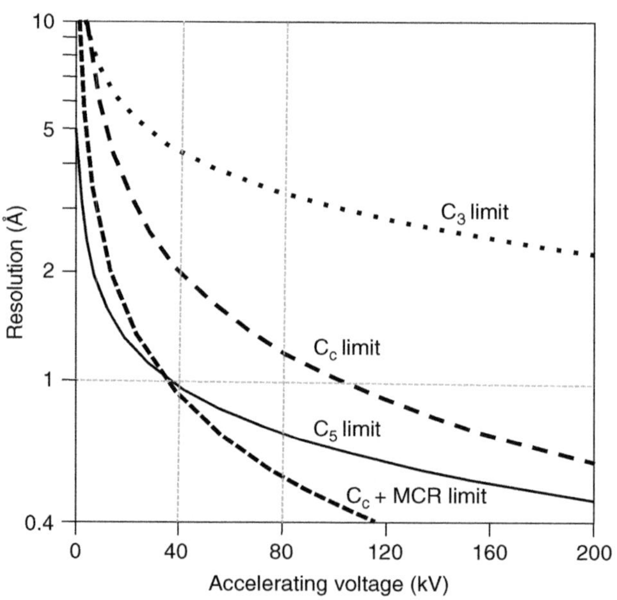

Figure 5.1 Resolution defined by the $1/e^2$ value of the contrast transfer function plotted as a function of electron energy for the aberration limits of: third order spherical aberration C_3, chromatic aberration C_c, fifth order spherical aberration C_5 and limit of the chromatic aberration with the energy spread reduced by the monochromator (C_c + MCR). (Reproduced with permission from Bell *et al.*, 2012).

energy-loss spectroscopy to determine band gaps and the dielectric properties with nanometer spatial resolution: the relativistic losses are reduced and delocalization of the energy-loss signal is reduced with the energy of the incident probe (Stoger-Pollach and Pongratz, 2010). Moreover, improvements in lower voltage (<80 keV) performance are anticipated to produce many new results in materials science because of reduced electron beam damage to sensitive samples (Meyer *et al.*, 2009).

The incorporation of aberration correctors, in particular the successful adoption of C_s correctors such as those by Corrected Electron Optics Systems (CEOS) in TEM columns has allowed the increased resolution now possible at sub-Angstrom levels (Haider *et al.*, 1995; Rose, 2009; Dahmen *et al.*, 2009). In the process of improving the resolution at high accelerating voltages, that is, above 200 keV, the C_s corrector has also enabled the improvement of resolution at lower energies (<120 keV) (Bell *et al.* 2010). Historically, the use of higher TEM voltages were favored since they reduce the spherical and chromatic aberration effects (O'Keefe *et al.*, 2001; Spence, 2003), but the development of spherical aberration correctors has allowed atomic resolution at 80 keV, 60 keV (Krivanek *et al.*, 2010; Sasaki *et al.*, 2010; Bell *et al.*, 2010; Kaiser *et al.*, 2011), and recently results showing atomic resolution at 40 keV have been reported (Bell *et al.*, 2012).

5.2 SO HOW LOW IS LOW?

Low voltage means different things to different people, and with this chapter we relate low voltage to mean lower than current "intermediate" TEM operating voltages. From the 1970s, historically speaking, there have always been TEMs that could operate at low voltages, below say 80 keV. In fact, an the beginning of TEM development 40 keV and 60 keV were chosen as a standard voltage by instruments from the Radio Corporation of America (RCA) and from Japan Electron Optics Laboratory (JEOL). When we say low voltage TEM, we mean the types of voltages that can be reached with a conventional TEM, although these could be modified with aberration correctors and other enhanced components for optimization of low keV performance.

What we describe here is completely different from low-energy electron microscopy, or LEEM, which is an analytical surface science technique invented by Ernst Bauer in 1962 (Bauer *et al.*, 1994, 1998). LEEM is a technique used to image atomically clean surfaces and thin (crystalline) films. In LEEM, high-energy electrons (15–20 keV) are emitted from an electron gun, focused using a set of optics, and sent through a magnetic beam deflector. The "fast" electrons travel through an objective lens and begin decelerating to low energies (1–100 eV) near the surface because the sample is held at a potential opposite that of the gun. The low-energy electrons are now "surface-sensitive" and the near-surface sampling depth can be varied by tuning the energy of the incident electrons (difference between the sample and gun potentials minus the work functions of the sample and system). The low-energy elastically backscattered electrons travel back through the condenser lens, reaccelerate to the gun voltage (because the condenser lens is grounded), and pass through the beam separator again. Imaging of the back focal plane of the objective lens into the object plane of the projector lens (using an intermediate lens) produces a diffraction pattern (low-energy electron diffraction, LEED) intensity distribution of the diffraction pattern will depend on the periodicity at the sample surface. Certain aspects of the techniques developed for LEEM can and will be transferred to low voltage TEM and STEM with further generations of instrumentation.

5.3 THE EFFECT OF CHROMATIC ABERRATION AND CHROMATIC ABERRATION CORRECTION

The term "chromatic aberration" comes from light optics: the light of different colors or wavelengths is focused with different focal lengths due to the dispersion of the glass used for the lenses. The magnitude and even the sign of the dispersion result from the energy (or wavelength) dependency of the interaction process of the photons with the lens material. Because there are materials with positive or negative dispersion, it is fairly easy to construct compound lenses that do not show chromatic aberration; an achromatic lens or achromat is a lens that is designed to limit the effects of chromatic and spherical aberration. Achromatic lenses are corrected to bring two optical wavelengths (typically red and blue which are the end limits of the visible spectrum) into focus in the same plane.

Electrons do not have colors, but their energy varies slightly about the mean beam energy. Since electron lenses are just electromagnetic fields, chromatic aberration describes the simple effect that faster electrons are deflected less by these fields than slower ones (see Chapter 1). From this fact it becomes clear that a single lens in electron optics always has positive chromatic aberration. One might think that a clever arrangement of several lenses might make it possible to compensate for this aberration. However, Scherzer proved in the same paper that this is never the case with round lenses (Scherzer, 1936).

There is an electron optical element, which has a freely tunable dispersion: the Wein filter that consists of mutually perpendicular electric and magnetic fields. One can extend this concept to mutually perpendicular electric and magnetic quadrupoles. Such an element is a focusing system with freely tunable dispersion, that is, some type of a lens with variable chromatic aberration. Unfortunately, a quadrupole – due to its two-fold symmetry around the optical axis – shows the negative dispersion, which can be used to compensate for the chromatic aberration of an objective lens, but only in one section. Therefore it does not help to use this element somewhere in a microscope column. The resultant correction optics require a multi-element corrector of mechanical and electronic complexity, further referenced in (Rose, 2009; Kabius *et al.*, 2009; Haider *et al.*, 1995). It is important to understand the limits to which chromatic aberration places on the ability to image especially at lower voltages. Chromatic aberration has two main effects: first, the image is blurred resulting in a loss of fine details and the contrast of the image is reduced. Secondly, the fact that due to chromatic aberrations, electrons with energies that differ from the mean beam energy form a wide and noisy background in the image caused by defocus spread.

Consequently, chromatic correction or the reduction of the effect of chromatic aberration would help to get clearer and crisper images with good image contrast. In transmission type instruments, the quality of the representation of fine details in the image is typically discussed in terms of the so-called damping envelope; this quantity describes how much the contrast of image details of a given size is reduced due to incoherent properties of the imaging instrument. Figure 5.2 shows the contribution of chromatic aberration to the damping envelope for several different accelerating voltages. Cc correction has several advantages as compared to monochromatization. Cc correction does not alter the properties of the electron beam source. Therefore, the brightness of the electron source is undiminished and the full beam current of a Schottky emitter ($\sim$100 nA) can be used, while a monochromatic system diminishes the beam current. Monochromatization improves the energy width ΔE by about an order of magnitude while Cc can be improved by three orders of magnitude to a few micrometers (Kabius *et al.*, 2009).

Some applications require the imaging of the specimen with a large energy width ΔE. One example is the imaging of biological specimens. Here, due to radiation damage, one would like to work with the minimum dose. Unfortunately, a lot of electrons that carry useful object

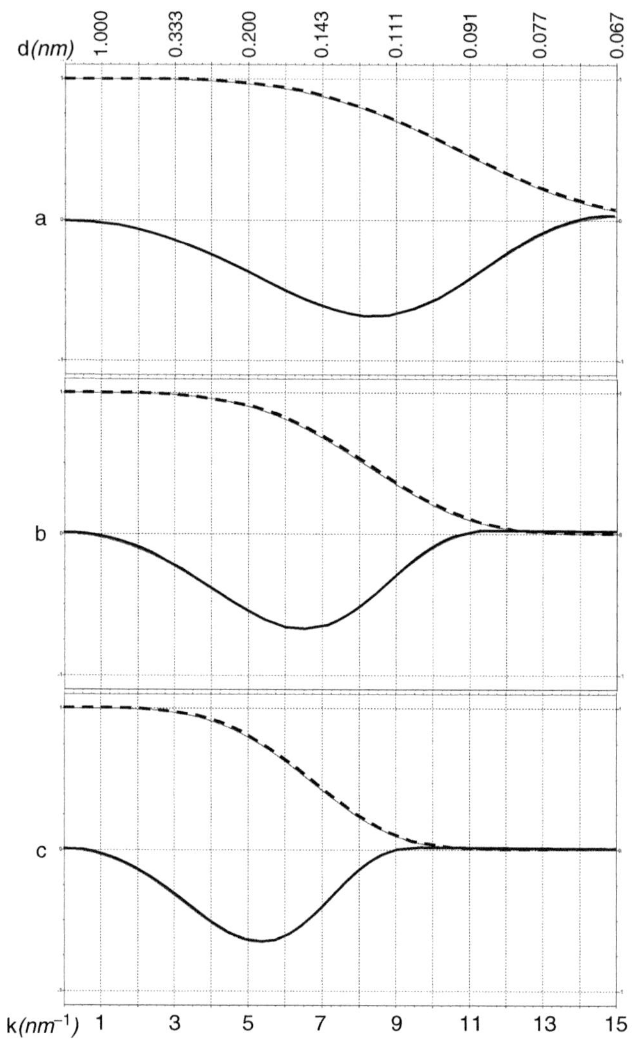

Figure 5.2 (a) Contrast Transfer Function of 200 keV Cs corrected transmission electron microscope (Zeiss Libra TEM); (b) CTF of 80 keV Cs TEM; (c) CTF for 40 keV.

information due to the elastic scattering within the specimen, also undergo some inelastic scattering while passing through the sample and the embedding substrate (ice, carbon film, etc.). Inelastic scattering increases the energy width of the beam far beyond the initial energy width of the electron gun of only about 0.6 eV. Another example is element specific imaging. In this technique, only those electrons are imaged that have experienced an energy loss characteristic for a certain element thus producing an element distribution image for this type of atom. To get a reasonable signal count, all electrons within the spectral line of the characteristic loss should be imaged. Therefore, the effective energy width becomes as large as the spectral line width, which is also significantly larger than the initial energy width of the gun. Taking as an example the Contrast Transfer Function of Cs corrected transmission electron microscope operated at 200 keV (Figure 5.2a), 80 keV (Figure 5.2b) and 40 keV (Figure 5.2c), the increase in resolution by the reduction of chromatic aberration is very obvious.

5.4 THE ELECTRON MONOCHROMATOR

Since chromatic aberration (C_c) is relatively expensive and difficult to correct, the resolution improvement by use of a monochromator is an effective option for materials applications. Until recently, lowering the primary beam energy meant limiting the resolution due to the limitation of the electron optics. Nowadays, the spherical aberration correction has become somewhat routine, thus making the chromatic aberration the limiting factor on achievable resolution, and particularly critical with lower keV imaging. Even without the benefit of a C_c corrector it is possible to reduce the effect of the chromatic aberration by using the electron monochromator to reduce the energy spread.

High resolution instruments (at typical intermediate voltages, 200–300 keV) have a very short focal length and a rather small chromatic aberration. For example the damping at 200 keV beam energy is comparatively low. Nevertheless, information about details on the order of 0.1 nm is almost entirely wiped out. As chromatic aberration is, in part from relativistic corrections, inversely proportional to the beam energy, the gain by chromatic correction increases with decreasing energy.

Figures 5.1 and 5.2 show that only some reduction of the effect chromatic aberration with an electron monochromator will provide sub-Angstrom (0.1 nm) resolution at 200 keV or 80 keV. Hence, for the first time, atomic resolution is available for specimens that are too sensitive to be observed at 300 keV, or which like light atoms, do not show sufficient contrast at such high beam energy.

Figure 5.1 shows the lower bounds to resolution imposed by spherical and chromatic aberration plotted for a Zeiss Libra 200 MC Cs corrected TEM. Without spherical aberration correction, the microscope is limited by the physically correctable third order spherical aberration C_3. If we define the information limit imposed by C_c as the spatial frequency at which the envelope damping function falls to a factor of $1/e^2$ times its maximum value, we can then write a simple formula to plot this as a function of electron energy limited by the spherical and chromatic aberration. From the well-known Scherzer defocus relation, we can write the lower bound in resolution, d_s, imposed by spherical aberration, C_3, which is the first zero crossing in the contrast transfer function at optimum defocus conditions (Scherzer, 1949):

$$d_s \geq \left(\frac{C_3\,\lambda^3}{6}\right)^{1/4} \tag{5.1}$$

For the resolution limit by chromatic aberration d_c we can use the expression from (Chang *et al.* 2006) and express it in terms of a change in energy spread. For the purposes of this discussion we ignore the contributions of the voltage and objective lens current to isolate the effect of the monochromator. A more exact calculation would require inclusion of these terms as well in the damping function, but we note that empirically we have found that these stabilities typically improve at lower voltages.

$$d_c = \sqrt{C_c \frac{\lambda^2}{2\sqrt{2}} \,{}^{\Delta E}\!/_{E}} \tag{5.2}$$

As shown by Figure 5.1, this will define a lower resolution limit bounded by the chromatic aberration C_c. With a 100 meV monochromated energy spread, which empirically provided the best compromise between beam brightness and reduced energy spread for the Libra 200MC instrument, this limit can be significantly reduced as shown in Figure 5.1 by the C_c + Monochromator (MCR) plot. Also from (Chang *et al.*, 2006) a lower limit bound by the fixed fifth order spherical aberration C_5 is defined as:

$$d_{C_5} = \left(C_5\lambda^5\right)^{\frac{1}{6}}/1.47 \tag{5.3}$$

and this limit is plotted for $C_5 = 1$ mm as the dashed line in Figure 5.1. The plot shows that at 40 keV the microscope is almost exactly at the cross over between the C_5 and C_c + MCR limit. Below 40 keV the monochromator provides a benefit, but the chromatic aberration still dominates; above 40 keV the limiting factor on Libra 200MC instrument is not chromatic aberration, but the fixed C_5 limit, thus instrumentally the obtainable resolution is limited at lower primary beam voltages by C_5.

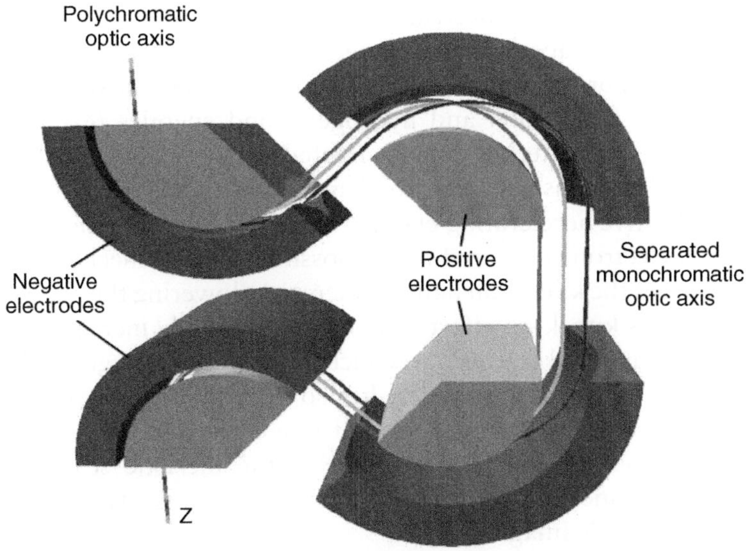

Figure 5.3 The Omega electron monochromator (CEOS) (Reproduced with permission from Bell et al., 2012) (See plate section for coloured version).

One way to provide electron monochromation besides a Wien filter is using an Omega type monochromator (Figure 5.3). The Omega-type monochromator is the only configuration that provides dispersion-free illumination, high brightness (108 A/cm^2 Sr) with non-distorting optics, as detailed in (Rose, 2009). This design is purely electrostatic and introduces no changes in source magnification during operation. The beam is deflected in an S-shape and forms an astigmatic image of the virtual source in the selection plane, which is the symmetry plane of the monochromator. There the beam is energy-filtered using an opaque slit, and an array of various slit sizes are available to allow optimization of energy width versus beam current. After leaving the monochromator, the beam is entirely dispersion-free; even the angular dispersion is compensated. It is also possible to switch the monochromator off and operate the microscope with a straight-through beam, as in ordinary operation mode.

5.5 THEORETICAL TRADEOFFS OF LOW KV IMAGING

In high resolution electron microscope imaging there are always tradeoffs between primary beam energy, contrast and damage. For any materials system it is possible to construct a plot of the relative atomic recoil energies at different accelerating voltages, (Bell *et al.*, 2012) and compare that to the bond breaking energies for the same components. An example of such a plot for carbon and hydrogen bond energies (exemplifying typical biological materials bonds) related bond energies is shown in Figure 5.4. This construction can then be used directly to determine the most effective accelerating energy to use to obtain the most efficient use of each electron compared to the possible damage mechanisms. For higher energies the knock-on damage increases; lowering the accelerating voltage reduces knock-on effects, but correspondingly increases the rate of radiolysis (inelastic scattering), which is a critical damage mechanism in biological and organic samples. It also occurs in inorganic samples, but is suppressed by the availability of free electrons. The tradeoff for imaging, of course, is that by increasing the inelastic scattering the contrast is also increased, usually a sought after quantity in biological and organic sample imaging.

By adjusting the microscope voltage to suit the sample, we can maximize the contrast to damage ratio while using aberration correction to maintain sufficient resolution to image the structure of interest. However, this is only possible if the microscope itself has a range of possible operating voltages and resolution can be maintained with each voltage. Most transmission electron microscopes operate only at fixed voltage steps, usually not low enough to gain possible contrast enhancements. The other matter to consider is the sample itself; lower voltages require thinner samples; this of course may or may not be possible to achieve depending on the material. Although the TEM samples must be significantly thinner for low keV observation, the improvement in contrast for inorganic materials, biological samples and especially nanobiological samples in low-voltage TEM while retaining atomic resolution cannot be understated. Damage mechanisms for biological samples are very complicated and very structure dependent; a comprehensive investigation of such systems is beyond the scope of this chapter.

Following on from the discussion in Chapter 1, Figure 5.5 illustrates a calculated angular distribution of elastic and inelastic scattering cross sections calculated in carbon at an accelerating voltage of both 40 and 100 keV showing how the scattering cross section decreases rapidly from zero degrees. Interestingly and of note is how the inelastic scattering starts higher than elastic and then drops off more rapidly than the elastic. Moreover, for the lower voltage of 40 keV the inelastic scattering cross-section shows it starts lower, increases and then matches the 100 keV. This effect is readily exploited to enable more contrast as expected

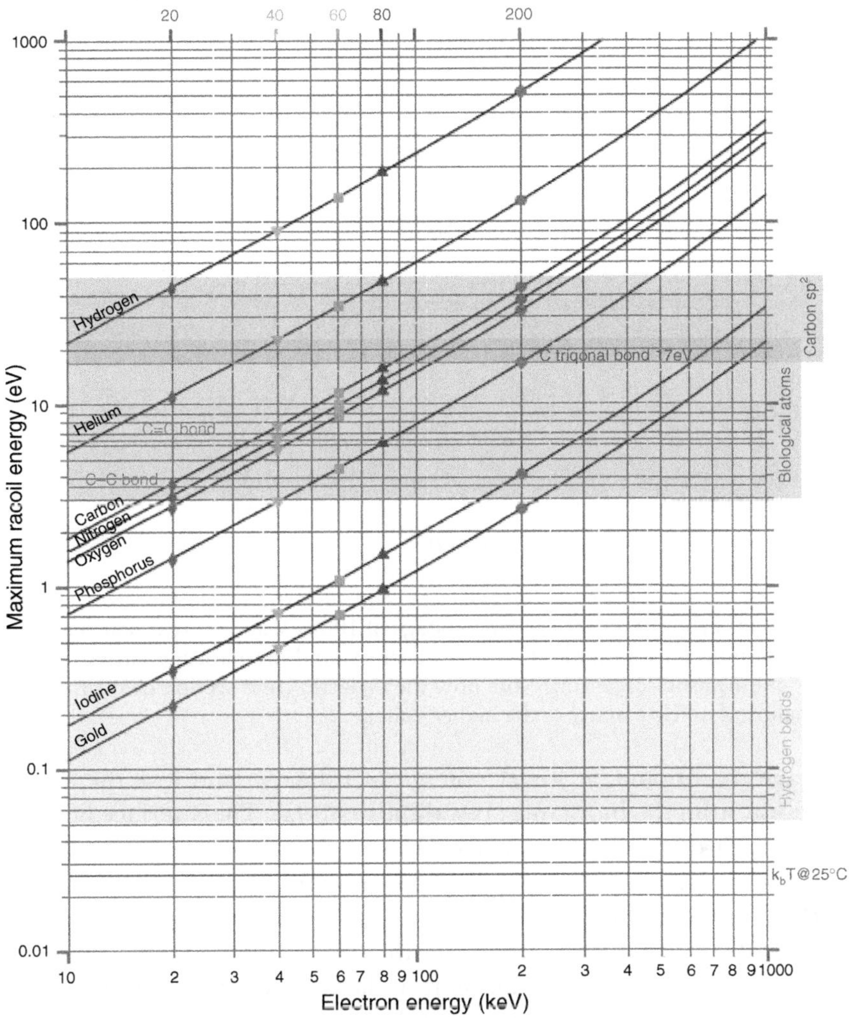

Figure 5.4 Maximum atomic recoil energy as a function of incident electron energy, colored regions indicate the bond breaking energies (Reproduced with permission from Bell *et al.*, 2012) (See plate section for coloured version).

for the lower voltages. For more detailed discussion and analysis of differential cross sections see, for example; Egerton (2011) and Reimer and Kohl (2008).

The imaging situation becomes even more interesting at lower voltages when an organic or biological material is coupled to an inorganic crystalline structure. Previously the choice would be to either image the inorganic at high energy (>200 keV) and destroy the organic material

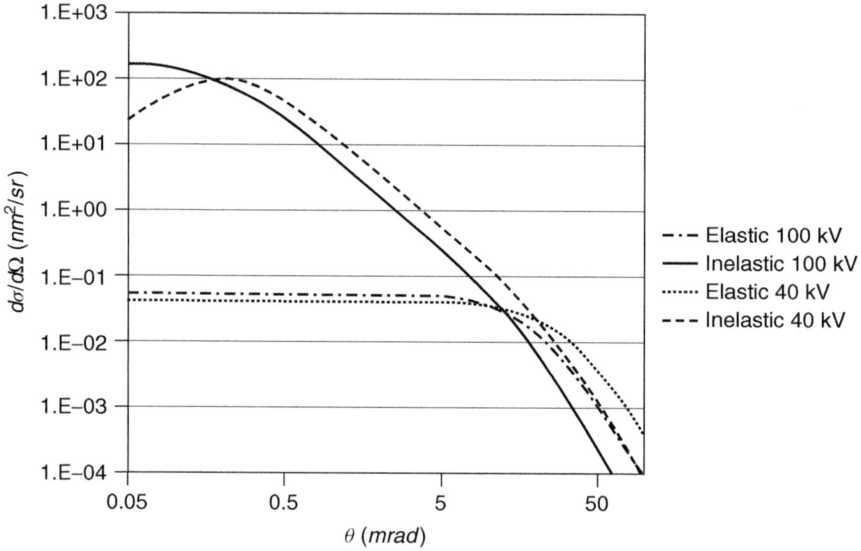

Figure 5.5 Angular distribution of elastic and inelastic scattering cross-sections calculated in carbon at an accelerating voltage of 100 keV and 40 keV. The scattering cross-section decreases rapidly from zero degrees, and dramatically varies between elastic and inelastic scattering. Note how the inelastic cross section changes at very low angle and quickly matches the higher voltage.

or image the organic at lower voltage (<100 keV) and lose the atomic resolution from the inorganic crystalline material. The interface between soft and hard materials is currently underexplored territory that is critical to underrated structure relationships with the hard/soft materials interface domain. You need high resolution of the hard material, but good contrast (and resolution) of the corresponding soft materials without inducing significant amount of damage to either component.

As an example, gold nanorods can be used as photothermal converters, permitting near-infrared (NIR) light to be transmitted deep into tissues without causing damage (Figure 5.6a). The temperature-sensitive material poly(N-isopropylacrylamide) (pNIPAm) collapses when heated beyond 32°C, and has been used in drug delivery systems (Timko *et al.*, 2011). Composites between these two disparate materials can be triggered with NIR light and open new avenues for triggerable drug delivery systems (Timko *et al.*, 2010). Toward this end, pNIPAm nanogel particles were synthesized and conjugated to polyelectrolyte-coated gold nanorods via electrostatic interactions (Karg *et al.*, 2007). LV-HRTEM enables imaging of both the polymer and gold components of these composite structures, and elucidates their interface (Figure 5.6b).

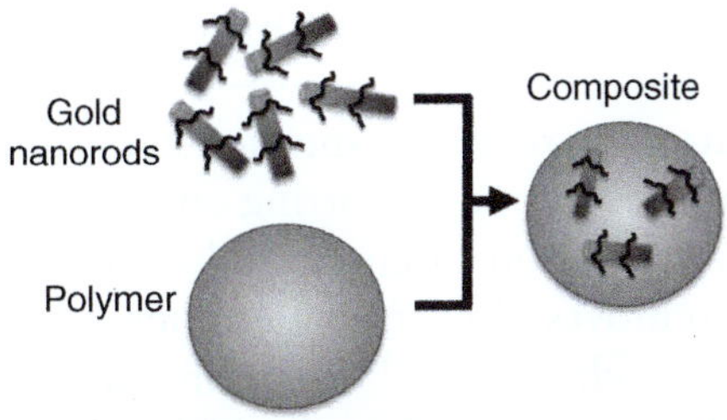

(a)

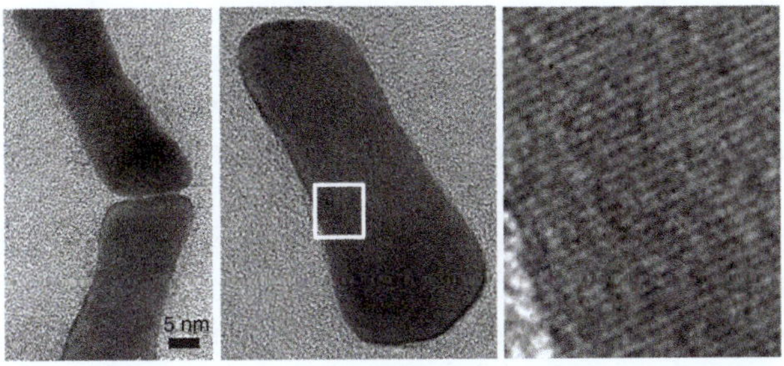

(b)

Figure 5.6 (a) Gold nanorods can be used as photothermal converters, permitting near-infrared (NIR) light to be transmitted deep into tissues without causing damage; (b) LVTEM images at 80 keV showing both the surrounding thermoplastic and the high resolution phase contrast lattice image of the gold.

5.6 OUR EXPERIENCE AT 40 KEV LV-HREM

Our group at Harvard has used for the majority of the examples published here a Zeiss Libra 200 monochromated, Cs-corrected transmission electron microscope (TEM) tuned to operate at 40 keV. This instrument has a Schottky-FEG with a maximum accelerating voltage of 200 keV, a third-order CEOS imaging corrector and an in-column, third generation Omega filter. The chromatic aberration coefficient, Cc, is approximately 1.2–1.3 mm, including the increase due to the corrector length and stability of the high voltage power supplies. For 40 keV imaging special attention was required to reduce the illumination tilt as low as possible due to increased deflector sensitivity and further adjustment using the current center (after Cs tuning) to again fine tune the tilt for the selected monochromator slit.

The post-specimen, mechanical beam blanker was used, which reduced instabilities in the sample induced by pulsed irradiation. For low-dose imaging, other blanking arrangements are possible. At 40 keV, the CEOS spherical aberration corrector requires more time than the microscope to stabilize; therefore, tuning adjustments were done in stages over three days (Bell *et al.*, 2012).

The monochromator provides dispersion-free illumination, high brightness (108 A/cm^2 Sr) and non-distorting optics as described previously. In the normal configuration of energy selection slits, the center slit is large (60 mm) and removes almost no electrons. Using the 60 mm slit is referred to within this chapter as "non-monochromated" operation, as this allows the full energy spread of the FEG to pass through to the accelerator chain. The in-column Omega energy filter electron optics are second order C_s corrected and third-order optimized, and show a high transmissivity (Tr = 190 nm^2 at 1 eV energy width), and a dispersion of 1.85 mm/eV at 200 keV with uniform isochromaticity in the imaging plane. A monochromated source and energy filtering for high-resolution imaging produces an increase in the cut-off of the temporal coherence damping envelope resulting from reduced chromatic aberration (Rose, 2009). Specifically, reducing the source energy distribution ($\Delta E/E$) increases the maximum spatial frequency of the envelope function (the information limit), thus improving the overall resolution, particularly for C_s corrected instrument.

5.7 EXAMPLES OF LV-HREM IMAGING

The resolution of the Libra 200MC TEM at 40 keV was initially tested with a specimen of gold nanoparticles sputtered onto ultra-thin (approximately 2 Å thick) carbon film. The resulting images are shown in Figure 5.7. Figure 5.7(a) shows an image of the sample obtained by utilizing only Cs corrector optics; the image shows no visible crystalline features and poor resolution. Figure 5.7b shows the same sample imaged using a C_s corrector coupled with a monochromator (1 μm slit), clearly resolving gold lattice fringes. As a comparison, when the same sample is imaged in this microscope at 80 keV, the images with and without the monochromator appear identical (Figure 5.8). The critical difference between 80 keV and 40 keV performance is that the relative contrast of the supporting carbon film is greatly enhanced in the 40 keV image due to the increased inelastic scattering (Figure 5.8a and b).

At 40 keV with a 0.5 mm monochromator slit, the electron beam energy distribution was measured to be FWHM = 43 meV; this measurement is limited by the dispersion resolution of the spectrometer and the actual value is likely slightly better. Previous results (Bell *et al.*, 2010) have shown that the 1 mm monochromator slit reduces the energy

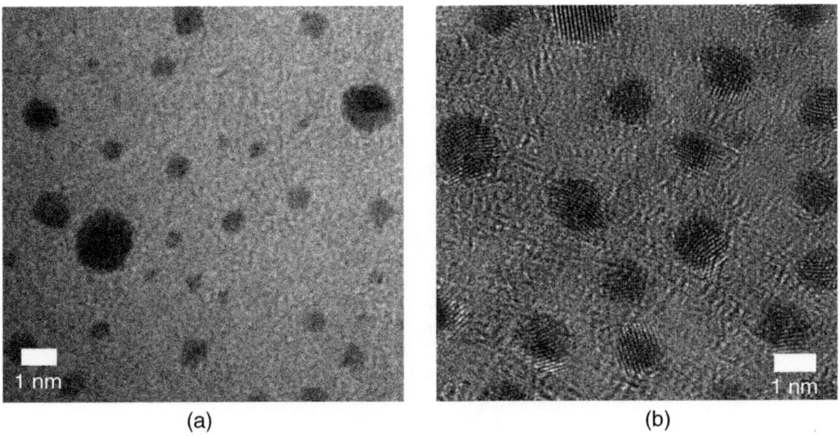

Figure 5.7 40 keV imaging of gold nanoparticles sputtered onto ultra-thin carbon film: (a) without the monochromator showed poor resolution with no observable lattice fringes; (b) with the monochromator with a 1 μm slit allows imaging of gold lattice fringes. (Reproduced with permission from Bell *et al.*, 2012).

spread of the electron beam to 81 meV at 200 keV and 62 meV at 80 keV, while still leaving sufficient current for usable high-resolution imaging (Figure 5.9). It was determined that for 40 keV imaging a 1 mm slit is optimal for typical use, resulting in energy spread of approximately 50 meV and a beam current of approximately 10 pA.

The point resolution was also measured by determining the separation between Si {004} dumbbell structures of 1.36 Å on a well-studied crystalline sample, Si <110>. The sample prepared by mechanical thinning and Ga FIB milling, with an additional step of Ar ion milling to remove the Ga beam damage; however, the sample was left with a slightly larger than expected amorphous edge, as is apparent in the image (Figure 5.10). The micrograph shows the general atomic structure of Si <110> with the Si {111} lattice planes clearly visible (3.14 Å). Close examination near the thin amorphous edge reveals the Si dumbbells (Figure 5.10, inset), {004} spacing measured at 1.36Å, which agrees with the accepted values (Kaiser *et al.*, 2011). The silicon dumbbell spacing is only apparent very close to the edge of the sample, with the range of suitable sample thickness for high resolution imaging being smaller than at higher electron beam energies. The sensitivity of electrons to being absorbed or scattered due to the silicon mass thickness is inversely proportional to the electron energy (Reimer and Kohl, 2008); for 40 keV electrons the inelastic mean free path in silicon is approximately 40 nm (Tanuma *et al.*, 1993; Ziaja *et al.*, 2006), which would indicate

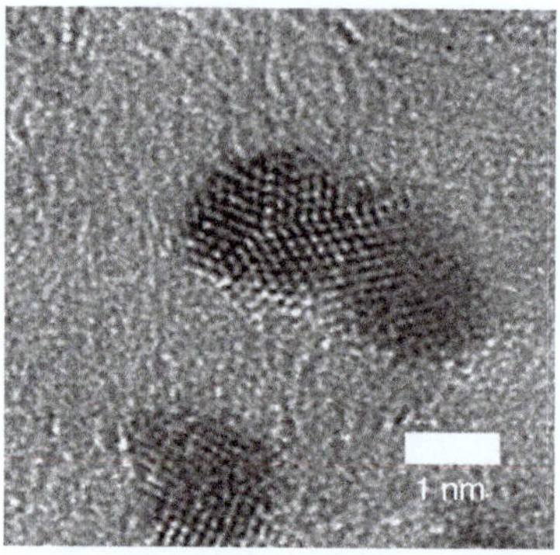

(a)

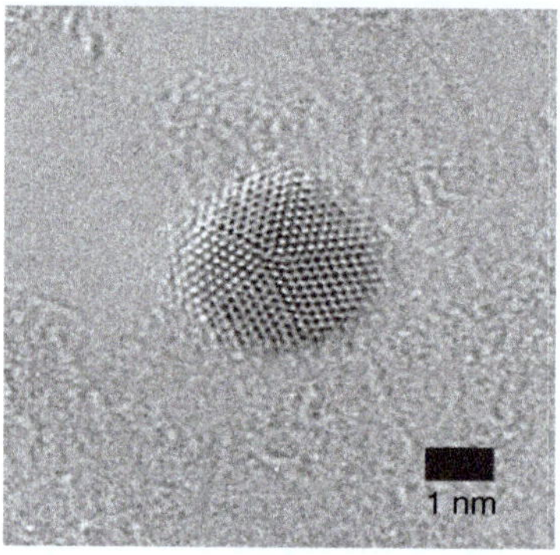

(b)

Figure 5.8 Gold nanoparticles on ultra thin carbon film: (a) Imaged at 40 keV with monochromator. (b) Imaged at 80 keV with monochromator (Reproduced with permission from Bell *et al.*, 2012).

that suitable samples should be less than about one electron mean free path thick for high resolution imaging for this keV regime. A multislice simulation (Kilaas, 2011) shows the importance of specimen thickness considerations for obtaining high-resolution images at low voltages, with the ideal sample required to be less than 100 Å thick.

Figure 5.11 shows an image of gold nanoparticle islands on a multilayer graphene film, which were previously unattainable at energies above 86 keV due to knock-on damage of the graphene substrate (Smith and Luzzi, 2001). Close examination of the images shows the underlying graphene film. Imaging such structures without damaging the graphene substrate may allow a new understanding of how the substrate interacts with the Au nanoclusters (van Huis *et al.*, 2009; Girit *et al.*, 2009). To obtain this image at 80 keV, the monochromator was adjusted to produce an energy spread of $\Delta E = 0.1$ eV, with the $C_s = -1.2\,\mu$m.

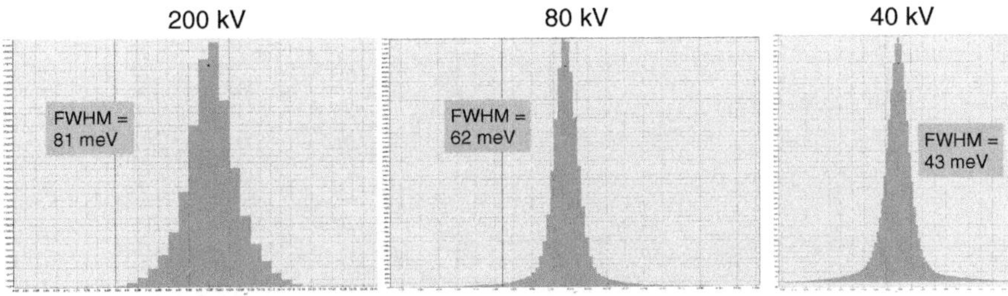

Figure 5.9 The measured electron energy spread with an omega electron monochromator at 200 keV, 80 keV and 40 keV with a 0.5 μm monochromator slit.

Figure 5.10 Si <110> at 40 keV. Inset edge detail of the Si <110> image showing Si dumbbell resolution of 1.3 Å (Reproduced with permission from Bell *et al.*, 2012).

The suspended graphene film can also be imaged directly at 80 keV (Figure 5.12). The advantage implied by Figure 5.5 is that at 40 keV the user is likely positioned below the bond breaking threshold for many materials. One good example is boron nitride that would be an interesting sample to study in the 40–60 keV range. Other materials include many atomically thin membranes recently described in (Coleman *et al.*, 2011). The image shown in Figure 5.12 is a single, one second exposure

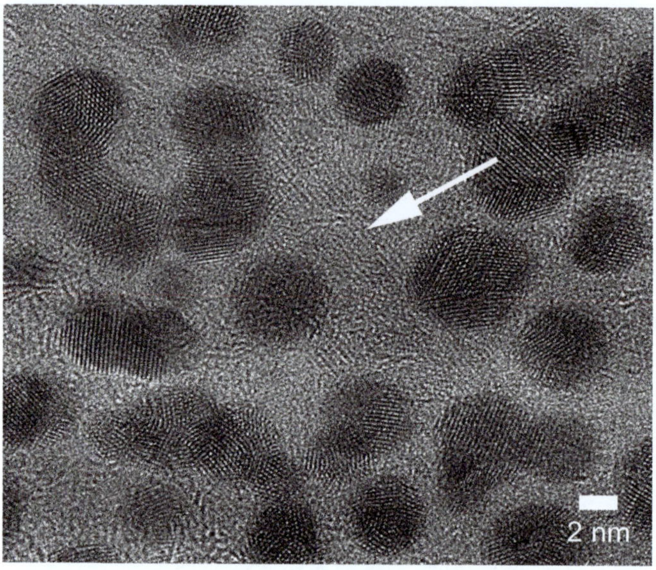

Figure 5.11 Gold nanoparticle islands on multi-layer graphene film imaged with electron monochromation ($\Delta E = 0.1$ eV) with Cs correction of $-1.2\,\mu$m and accelerating voltage of 80 keV. The Au fringes and the underlying graphene structure are clearly visible in the indicated areas between the nanoparticles (arrow). (Reproduced with permission from *Microscopy and Microanalysis*; Bell *et al.*, 2011).

of a multilayer graphene membrane, low pass filtered to 3 pixels (0.70 Å) with a 5% roll off.

Another example of atomic resolution at 80 keV is shown in Figure 5.13. The sample is a cluster of single-wall carbon nanotubes suspended over a hole in a carbon membrane. The image clearly shows the carbon atom arrangements in the sidewalls of the nanotubes (Suenaga *et al.*, 2007). Such images would not be possible without the reduced damage and the Ångstrom resolution of the Cs/MC instrument operating at 80 keV. This image was obtained using zero-loss energy filtering at 80 keV with Cs correction and a monochromator slit width of 4 µm.

5.8 CONCLUSIONS

Today a TEM equipped with a spherical aberration correction and monochromator is able to image at atomic resolution at lower accelerating voltages. The direct advantage of lowering the beam energy is the increased inelastic scattering cross section that improves contrast for imaging, but at higher resolution than previously possible.

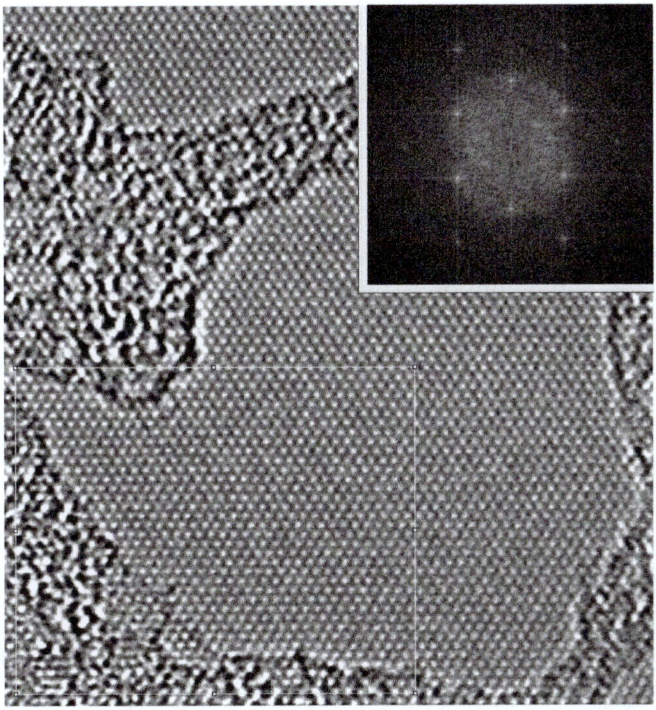

Figure 5.12 Graphene imaged at 80 keV with 3 μm MC slit ($\Delta E = 230$ meV), Inset FFT of graphene lattice.

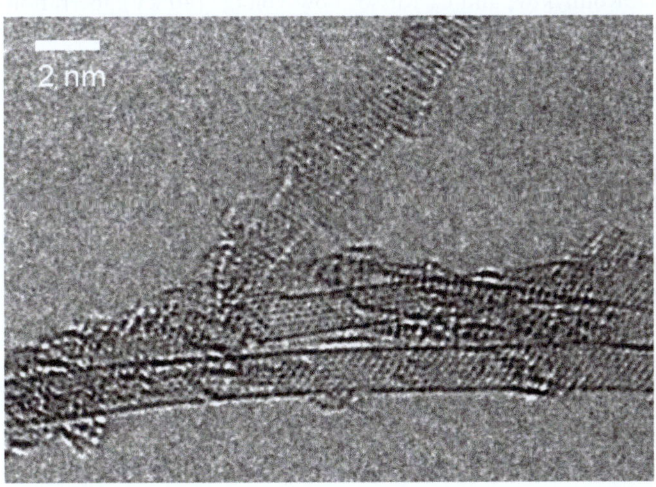

Figure 5.13 Single wall carbon nanotubes imaged at 80 keV with C_s correction, a monochromator slit width of 4 μm, and zero-loss imaging (Reproduced with permission from *Microscopy and Microanalysis*, Bell *et al.*, 2011).

In the low energy regime the limiting factor for imaging resolution becomes the chromatic aberration; by reducing the energy spread with the monochromator the effects of the chromatic aberration are reduced and the resolution is improved. This has immediate impact on the imaging of the biological/materials interface where phase contrast and scattering contrast are equally important.

Using the spherical aberration corrector in conjunction with electron monochromator for example at 40 keV takes the user surprisingly close to the lower bound imposed by fifth-order spherical aberration, and enables imaging with an information limit better than 1 Å, and a practical resolution of better than 1.4 Å. Low voltage operation will likely become the new mode of imaging for many electron microscopes, with the instrument being, in essence, tuned to extract all the information possible from each electron that transits the sample.

REFERENCES

E. Bauer, Low energy electron microscopy, *Rep. Prog. Phys.*, **57**, 895–938 (1994).

E. Bauer, LEEM basics, *Surf. Rev. Lett.*, **5**, 1275–1286 (1998).

D.C. Bell, C.J. Russo, and G. Benner, Sub-Angstrom low-voltage performance of a monochromated, aberration-corrected transmission electron microscope, *Microsc. Microanal.*, **16**, 386 (2010).

D.C. Bell, D. Kolmykov, and C. Russo, Low-voltage (40 kV) aberration-corrected, monochromated imaging for carbon nanostructures, *Microsc. and Microanal.*, **17**, 1490–1491 (2011).

D.C. Bell, C.J. Russo, and D.V. Kolmykov, 40 keV atomic resolution TEM, *Ultramicroscopy*, **114**, 31–37 (2012).

L.Y. Chang, A.I. Kirkland, J.M. Titchmarsh, On the importance of fifth-order spherical aberration for a fully corrected electron microscope, *Ultramicroscopy*, **106**, 301 (2006).

J.N. Coleman, M. Lotya, A. O'Neill, *et al.*, Two-dimensional nanosheets produced by liquid exfoliation of layered materials, *Science*, **331**, 568 (2011).

U. Dahmen, R. Erni, V. Radmilovic, *et al.*, Background, status and future of the Transmission Electron Aberration-corrected Microscope project, *Phil. Trans. R Soc. A.*, **367**, 3795–3808 (2009).

R.F. Egerton, *Electron Energy Loss Spectroscopy in the Electron Microscope*, 3rd edn, Plenum Press, New York (2011).

Ç.Ö. Girit, J.C. Meyer, R. Erni, *et al.*, Graphene at the edge: stability and dynamics, *Science*, **323**, 1705–1708 (2009).

M. Haider, G. Braunshausen, and E. Schwan, Correction of the spherical-aberration of a 200-kV TEM by means of a hexapole-corrector, *Optik*, **99**, 167 (1995).

P.W. Hawkes and J.C.H. Spence. *Science of Microscopy*. Springer; New York (2007).

B. Kabius, P. Hartel, M. Haider, *et al.*, First application of Cc-corrected imaging for high-resolution and energy-filtered TEM, *J. Electr. Microsc.*, **1** (2009).

U. Kaiser, J. Biskupek, J.C. Meyer, *et al.*, Transmission electron microscopy at 20kV for imaging and spectroscopy, *Ultramicroscopy*, **111**, 1239–1246 (2011).

M. Karg, I. Pastoriza-Santos, J. Pérez-Juste, *et al.*, Nanorod-coated PNIPAM microgels: thermoresponsive optical properties, *Small*, **3**, 1222–1229 (2007).

R. Kilaas, MacTempasX. Multislice Image Simulation Software (2011).

O.L. Krivanek, N. Dellby, M.F. Murfitt, *et al.*, Gentle STEM: ADF imaging and EELS at low primary energies, *Ultramicroscopy*, **110**, 935–945 (2010).

J.C. Meyer, A. Chuvilin, G. Algara-Siller, *et al.*, Selective sputtering and atomic resolution imaging of atomically thin boron nitride membranes, *Nano Lett.*, **9**, 2683–2689 (2009).

M.A. O'Keefe, C.J. Hetherington, Y.C. Wang, *et al.*, Sub-Angstrom high-resolution transmission electron microscopy at 300 keV, *Ultramicroscopy*, **89**, 215 (2001).

L. Reimer and H. Kohl, *Transmission electron microscopy: physics of imaging formation*, 5th edn, Springer, Berlin (2008).

H. Rose, *Geometrical Charged Particle Optics*, Springer, Berlin, Heidelberg (2009).

T. Sasaki, H. Sawada, F. Hosokawa, *et al.*, Performance of low-voltage STEM/TEM with delta corrector and cold FEG, *J. Electron Microsc.*, **59**, S7–S13 (2010).

O. Scherzer, Über einige Fehler von Elektronenlinsen, *Z. Physik*, **101**, 593 (1936).

O. Scherzer, The theoretical resolution limit of the electron microscope, *J. Appl. Phys.*, **20**, 20 (1949).

B.W. Smith and D.E. Luzzi, Electron irradiation effects in single wall carbon nanotubes. *J Appl Phys.*, **90**, 3509–3515 (2001).

J.C.H. Spence, *High-Resolution Electron Microscopy*, 3rd edn, Oxford University Press (2003).

M. Stöger-Pollach and P. Pongratz, Advantages of low beam energies in a TEM for valence EELS, *J. Phys. Conf. Series*, **209**, 012031 (2010).

K. Suenaga, H. Wakabayashi, M. Koshino, *et al.*, Imaging active topological defects in carbon nanotubes, *Nature Nanotech.*, **2**, 358–360 (2007).

S. Tanuma, C.J. Powel, and D.R. Penn, Calculation of electron inelastic mean free paths, *Surf. Interf. Anal.*, **20**, 77–89 (1993).

B.P. Timko, T. Dvir and D.S. Kohane, Remotely triggerable drug delivery systems, *Adv. Mater.*, **22**, 4925–4943 (2010).

B.P. Timko, K. Whitehead, W. Gao, *et al.*, Advances in drug delivery, *Ann. Rev. Mater. Res.*, **41**, 1–20 (2011).

M.A. van Huis, N.P. Young, G. Pandraud, *et al.*, Atomic imaging of phase transitions and morphology transformations in nanocrystals, *Adv. Mater.*, **21**, 4992–4995 (2009).

B. Ziaja, R.A. London, and J. Hajdu, Ionization by impact electrons in solids: electron mean free path fitted over a wide energy range, *J. Appl. Phys.*, **99**, 033514 (2006).

6

Gentle STEM of Single Atoms: Low keV Imaging and Analysis at Ultimate Detection Limits

Ondrej L. Krivanek[1], Wu Zhou[2,3], Matthew F. Chisholm[3], Juan Carlos Idrobo[2,3], Tracy C. Lovejoy[1], Quentin M. Ramasse[4] and Niklas Dellby[1]

[1]*Nion Co., USA*
[2]*Department of Physics and Astronomy, Vanderbilt University, USA*
[3]*Materials Science and Technology Division, Oak Ridge National Laboratory, USA*
[4]*SuperSTEM Laboratory, STFC Daresbury, UK*

6.1 INTRODUCTION

Scanning transmission electron microscopy (STEM), introduced in its modern form by Crewe and coworkers over 40 years ago [1–4], has made important advances in the last few years. Chief among these was the successful implementation of working aberration correctors, which has improved the available spatial resolution significantly: 1 Å is now routine and 0.5 Å is attainable. Single atoms as light as boron have been imaged using annular dark field modes and as light as hydrogen using annular bright field modes. Electron energy loss spectroscopy (EELS) and energy-dispersive X-ray spectroscopy (EDXS) have become standard techniques for analyzing the chemical composition of individual atomic

Low Voltage Electron Microscopy: Principles and Applications, First Edition.
Edited by David C. Bell and Natasha Erdman.

columns in crystals, and EELS and EDXS data from single atoms have been acquired. Most of these developments have been reviewed in a recent 760-page book on STEM that comprises 18 chapters on topics ranging from the history of STEM to new STEM techniques such as fluctuation imaging [5]. More recent developments such as single atom EDXS and fine structure analysis of EEL spectra acquired from single atoms are covered in this chapter.

Achieving atomic resolution at primary energies low enough to stay below knock-on damage thresholds in low Z materials such as graphene, nanotubes, and monolayer boron nitride has been an important part of the progress. Depending on which features in these materials are being studied, the knock-on thresholds generally lie between less than 10 keV (for adatoms loosely attached to a graphene sheet) and 85 keV (for perfect graphene monolayers). Aberration-corrected STEMs using cold field emission guns (CFEGs) can reach probe sizes of about 1 Å at primary energy of 60 keV and only slightly bigger than 1 Å at 30 and 40 keV [6, 7]. This allows individual atoms to be resolved even in densely-spaced materials such as graphene.

Maintaining ultra-high vacuum (UHV, pressure $<1 \times 10^{-9}$ Torr) at the sample, which minimizes sample contamination and also sample etching due to contaminants such as water being adsorbed and ionized into reactive species by the incident beam, has been another important development [7, 8]. Under UHV conditions, the absence of radiation damage in materials such as graphene, which does not suffer from ionization damage, allows large irradiation doses of the order of 10^7–10^{10} electrons/Å to be used. Doses of about 10^7 electrons/Å lead to STEM annual dark field (ADF) images in which the intensities of the images of individual atoms vary by only a few percent. ADF images in which the shot noise has been reduced to this low a level allow the chemical type of well-resolved, non-overlapping atoms to be identified simply by the intensity of their image [9]. Doses of 10^8–10^{10} electrons/Å^2 allow the chemical type of individual atoms to be determined spectroscopically, by EELS and EDXS [10–15]. Doses in the upper half of this range can lead to EELS spectra displaying fine structures that allow the atomic bonding of individual atoms to be investigated [14].

Because imaging samples at low energies can be nearly damage-free, we have taken to calling low energy STEM operation [11]: "gentle STEM". The technique joins the trend that has been important in scanning electron microscopy (SEM) for some time: lowering the operating energy. Further benefits of lower energy operation in the STEM are that the interaction between slower electrons and the sample is stronger (the cross-sections are larger), which increases the signal from each individual atom, and probably also that the EELS and EDXS signals are slightly better localized. These benefits are closely related to the primary reason for low-energy SEM, which is to decrease the interaction volume of the electron beam in the sample [16] in order to obtain higher-resolution information about the sample while producing less radiation damage. The STEM approach is the most valuable when it provides atomic resolution, and this restricts it to operating at energies higher than those commonly used in SEM. Nevertheless, a convergence is occurring between the two techniques. For instance, single heavy atoms have been imaged using the secondary electron (SE) signal in a medium-energy, aberration-corrected STEM [17].

We have described the low-energy operation of a STEM previously, particularly in References [11] and [18]. In this chapter we provide an overview of the principal parameters that need to be optimized to reach STEM atomic resolution at low primary energies, showcase recent examples of gentle STEM applications to single atom imaging and spectroscopy, and speculate about future developments.

6.2 OPTIMIZING STEM RESOLUTION AND PROBE CURRENT AT LOW PRIMARY ENERGIES

Three principal instrumental aspects need to be optimized in high resolution STEM: the brightness of the electron source, the optical performance of the probe-forming column, and the performance of the STEM detectors. A full description of the needed optimizations has been given in Reference [18]. Here we give a simplified account that is more focused on reaching the best performance.

Optimizing the performance of low (10–80) keV STEM is no different in principle from optimizing medium (100–400) keV STEM, but there are subtle differences. First, chromatic aberration tends to be a more severe limit on the attainable resolution at low operating energies than at higher ones, and extra attention therefore needs to be paid to it. Second, because the attainable resolution is not much higher than the inter-atomic distances that need to be resolved, geometric aberrations and other imaging defects need to be controlled more precisely than at higher energies. Third, not much beam broadening due to a finite source size can be tolerated, and this puts strict requirements on the quality of the electron source and on how much current can be contained in the probe.

The strategy for achieving the smallest possible probe in a given instrument is much the same for both aberration-corrected and uncorrected STEM: one chooses a condenser lens regime that provides a source demagnification value such that the probe current is sufficient for the task at hand, selects the "desirable part" of the electron beam by a beam-defining aperture, and fine-tunes the microscope optics. The aperture is usually called "virtual objective aperture" in dedicated STEM instruments, and "condenser aperture" in TEM/STEM ones. The "desirable part" of the beam is the part that is not seriously affected by aberrations, and will therefore contribute to the probe maximum rather than the probe tail.

The effect of geometric aberrations is easily assessed by a visual inspection of the shadow image of the sample called the electron Ronchigram [19–21]. Parts outside the featureless Ronchigram "sweet spot" that is obtained at the exact focus needed to be cut out by the aperture. Furthermore, automated aberration-tuning algorithms typically advise whether the measured aberrations are compatible with the aperture size being used or not. The effect of chromatic aberration is not easily assessed from a regular Ronchigram. The optimum aperture size to be used when chromatic aberration is the dominant limit and the resultant resolution therefore need to be computed from the values of the two parameters that determine the magnitude of the aberration: the chromatic aberration coefficient for the entire probe-forming column C_c and the energy spread of the electron beam δE. The energy spread arises due to both the natural energy width of the electron source and instabilities in the microscope high voltage. It is readily measured if an electron energy-loss spectrometer is a part of the STEM system. C_c is determined simply by changing the high voltage by ΔV and measuring the change in defocus ΔZ, and then using:

$$C_c = \Delta Z / \left(\Delta V / V_o \right), \tag{6.1}$$

where V_o is the primary voltage.

The optimum illumination half-angle α_o to be used when chromatic aberration is the dominant limit on the resolution is well approximated by [18]:

$$\alpha_o = \alpha_{chrom} = 1.2 \ \left(\lambda / \left(C_c \delta E / E_o \right) \right)^{1/2} \tag{6.2}$$

where λ is the electron wavelength, δE is the energy spread of the beam specified as full-width at half-maximum (FWHM), and E_o is the primary energy.

The FWHM of the diffraction-limited probe d_o for an illumination half-angle α_{chrom} is given by:

$$d_o = 0.61 \lambda / \alpha_{chrom} \tag{6.3}$$

and this leads to the diffraction-limited probe size $d_{o\ chrom}$ when the chromatic aberration is the major limit:

$$d_{\mathrm{o\ chrom}} = 0.5\ (\lambda C_{\mathrm{c}} \delta E / E_{\mathrm{o}})^{1/2} \tag{6.4}$$

The probe maximum can usually be adequately modeled by a Gaussian curve, even though the probe tail is usually much more pronounced than for the Gaussian shape. Because the addition of two identical Gaussian curves shifted by one FWHM gives a curve with a small dip (7%) halfway between the Gaussians, the probe FWHM value can also be taken as approximately the "resolution" of the image of an infinitely sharp object. Moreover, since the central portion, up to the half-maximum width, of a two-dimensional, rotationally symmetric probe with a Gaussian profile contains 51% of the probe's total intensity, for a Gaussian probe the FWHM width is also essentially the probe diameter containing 50% of the electrons (d_{50}).

Equation (6.4) gives the probe size in the limit of zero probe current; the broadening due to a finite current is discussed next. Note also that because λ is proportional to $1/\sqrt{E_0}$ at low energies (and to $1/E_o$ at fully relativistic energies), $d_{o\ chrom}$ has a strong inverse dependence on the primary energy. In other words, attaining good spatial resolution at low primary energies in the presence of chromatic aberration is much more challenging than at high ones.

The two parameters in (6.4) that are different for different instruments are the energy spread δE and the coefficient of chromatic aberration C_c. δE ~0.3 to 0.4 eV for an optimized cold field emission electron source operating with a total extraction current of a few μA. C_c ~0.8–1.0 mm for an optimized STEM column using an objective lens with an ultra-small gap (~2 mm), or a small gap lens (~ 4 mm) in the so-called two-crossover regime [7], plus condenser lenses and an aberration corrector optimized to give low C_c contributions. The "chromatic product" C_c δE can then be as low as about 0.25 mm eV. Under more regular conditions, the product can easily reach 0.5 mm eV even for cold field emission STEM, and 1 mm eV and higher for STEMs using Schottky sources with δE ~0.7 eV and C_c > 1.5 mm.

The next resolution-limiting influence to consider is the fact that at finite probe currents, the finite size d_{source} of the source projected into the sample plane makes the probe grow broader. The size of the broadened probe d_p can be approximated by:

$$d_p = \left(d_o^2 + d_{source}^2\right)^{1/2} \tag{6.5}$$

Using the fact that the current in the probe is proportional to the area of the source projected into the sample plane, that is, to d^2_{source}, we can rearrange (6.5) into a form that is more easily related to the way the microscope is run:

$$d_p = d_o \left(1 + I_p/I_c\right)^{1/2} \tag{6.6a}$$

$$d_p = 0.5\,(\lambda C_c \delta E/E_o)^{1/2} \left(1 + I_p/I_c\right)^{1/2} \tag{6.6b}$$

where I_p is the probe current, easily determined by a Faraday cage or any other device able to measure the probe current, and I_c is the so-called "coherent current" [18], defined as the probe current value for which $d_{source} = d_o$ and the probe broadens to $d_p = \sqrt{2}\, d_o$. Equations (6.6) describe how the probe size will broaden as the probe current is increased in any STEM (or SEM) operated at any primary energy, provided that the coherent current is known.

The coherent current I_c is a characteristic property of the source. It does not depend on the accelerating voltage (provided that stochastic Coulomb interactions that occur after the first anode can be neglected), or on the quality of the microscope's optics. In other words, with the same electron gun operated in the same way, the same coherent current applies to both aberration-corrected and non-corrected instruments, and to any accelerating voltage.

As shown in Reference [18], the coherent current is proportional to the normalized source brightness B_n:

$$|I_c| = 138\ B_n \tag{6.7}$$

where I_c is given in pA and the normalized brightness B_n in pA/(Å^2 sr V). Choosing these units makes the typical CFEG normalized brightness come out as ~1.

I_c is expected to lie between 100–500 pA for well-functioning cold field emission sources, and between 20–100 pA for Schottky sources. Unfortunately, I_c (or B_n) values are not readily available for any particular STEM. For CFE guns, the maximum attainable value of I_c is typically limited by stochastic Coulomb interactions within the electron beam, and these interactions depend on many parameters: the source workfunction, the source radius, the total emission current from the source, the forward percentage of the emitted beam, and where the beam crossovers and current-limiting apertures are. Even with most of these parameters held constant, the coherent current can vary from day to day. For instance, a contaminated cold field emission tip can give a significantly smaller I_c than a clean one. I_c can also vary over the long term, as the field emission tip changes its shape, and it can be degraded by high-frequency instabilities that broaden the source size. Getting an accurate measure of I_c is thus only possible if it is determined experimentally for each particular STEM instrument.

I_c is not difficult to determine: one simply measures the probe size d_p for a range of values of the probe current I_p, while running the gun in as constant a condition as possible. The different I_p values are obtained by varying the demagnification of the electron source projected onto the sample, that is, by changing the magnification setting of the condenser lenses. Rearranging Equation (6.6a) into:

$$d_p^2 = \left(d_o^2/I_c\right)\ I_p + d_o^2 \tag{6.8}$$

makes it clear that plotting d_p^2 vs. I_p should result in a straight line with slope d_o^2/I_c and intercept d_o^2, which means that I_c can be computed simply as $I_c = intercept/slope$. Proceeding this way allows the error of the determination to be estimated.

As with all practical measurements, there are important caveats. The probe size needs to be only diffraction and source-size limited for all the data recorded: if aberrations determine the probe size directly, by distorting the wavefront allowed through the aperture, the above approach will not work. The measurements need to extend to probe current values $I_p > I_c$ if they are to be reasonably accurate. Finally, a method for estimating the probe size (FWHM, not the highest spatial frequency that gives non-zero information transfer) must be used. In practice these requirements mean that the resolution should be determined for a series of probe currents from small (30 pA or less) to large (ideally 1 nA or more for a CFEG system), with about a 2 × smaller beam-defining aperture than normally used. The first point will guarantee that a range of I_p values that determines I_c accurately will be used, the second one that aberrations that are likely to grow larger at large probe currents will have little influence on probe formation. The second point is important because a larger probe current normally means a broader beam in the gun and the condensers, and this results in larger aberration contributions from this part of the column to the total aberrations of the system than when the probe current is small. If the size of the beam-defining aperture is chosen sufficiently small, the larger aberrations will not affect the shape of the probe in any substantial way, and will therefore not invalidate the measurement.

Several methods are available for estimating the FWHM size of the probe. One is to image stationary single heavy atoms [22] and to estimate the probe size from the width of the atomic images, using a statistically significant sample rather than just a few best-looking atomic images. Another method is to qualitatively compare images of crystalline materials with a precise simulation [23]. The probe size can also be worked out by determining quantitatively how much the magnitude of the spatial frequency spectrum of the imaged object has been reduced by the imaging, for instance by comparing the intensities of the Fourier transform (FT) of a recorded image with the intensities measured in a diffraction pattern recorded from the same sample. The source size and the overall probe size can also be estimated from the transfer strength of interference fringes in a Ronchigram of a crystalline sample [24]. The safest approach is of course to determine the probe size using at least two different methods, and to make sure that the results are similar.

Returning to the subject of the attainable resolution of the entire STEM system, Figure 6.1 shows the probe size expected in a well-tuned C_c-limited STEM for primary energies between 20–80 keV, with three different values of $C_c\ \delta E$:

1. 0.25 mm eV, which corresponds to a somewhat optimistic combination of $C_c = 0.8$ mm and $\delta E = 0.31$ eV;
2. 0.5 mm eV, corresponding to a more realistic $C_c = 1.25$ mm and $\delta E = 0.4$ eV; and
3. 1.0 mm eV, corresponding for instance to $C_c = 1.4$ mm and $\delta E = 0.71$ eV.

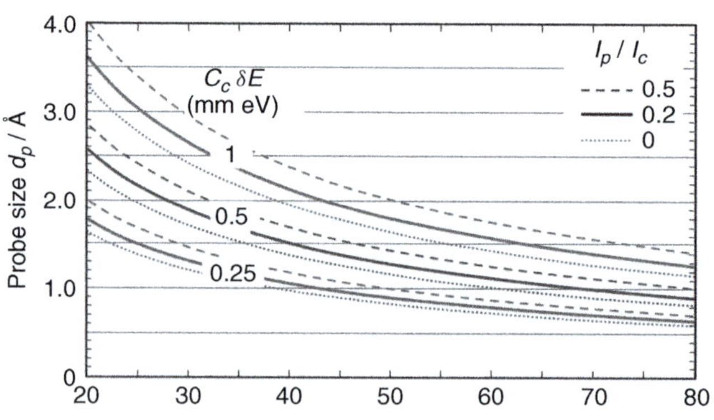

Figure 6.1 The dependence of the probe size d_p on the primary energy for three values of the chromatic product $C_c\ \delta E$ and three different ratios I_p/I_c of the actual beam current to the coherent current (See plate section for coloured version).

In addition, three curves are shown for each type of STEM: for $I_p = 0$, $I_p = 0.2\ I_c$ ($I_p = 60$ pA when $I_c = 300$ pA), and $I_p = 0.5\ I_c$. The figure makes it clear that the attainable probe size d_p is strongly dependent on the primary energy E_o, and that at larger probe currents, it also becomes strongly dependent on the probe current used. A 1 Å probe FWHM is attainable at about 40 keV for the optimistic estimate of $C_c\ \delta E$ and a negligible probe current, and at 60 keV for a readily attainable $C_c\ \delta E$ and also a negligible beam current.

Note that the probe FWHM is a much stricter criterion of the probe size than the "information transfer limit", usually determined as "the furthest reflection visible in a Fourier Transform, no matter how weak". For a Gaussian damping envelope, there is no absolute upper limit on how far the information transfer extends, and weak transfer of strong reflections can give rise to FT spots at twice the spatial frequency corresponding to the FWHM. One should therefore not confuse the presence of a spot in the Fourier transform with the probe FWHM, or with the resolution demonstrated by resolving two objects a certain distance apart.

Another important point is that at 100 keV and higher, higher-order aberrations even in C_5-corrected microscopes begin to make a non-negligible contribution to the resolution limit (see Figure 15.3 in Ref. [18]), and the simple model presented here, in which only the chromatic aberration is considered, ceases to apply. But at <100 keV, the simple model agrees well with experimental results, as for instance shown, further on in this chapter, by unprocessed ADF images of graphene that demonstrate clear separation of the images of the nearest neighbor C atoms. They illustrate that experimental resolution (and probe FWHM) better than 1.4 Å is readily attained at 60 keV.

The take home-lesson from the above considerations is that optimizing the set-up of a given STEM instrument for low energy imaging or spectroscopic analysis can and should be done in a systematic way. Provided that the geometric aberrations are sufficiently under control, and that there are no major resolution-destroying disturbances (such as sample vibrations or strong magnetic stray fields shaking the probe), knowledge of just five microscope parameters is needed to determine the probe size: the primary energy E_o; the chromatic aberration coefficient of the probe-forming optics C_c; the energy spread δE; the coherent current I_c; and the actual probe current I_p.

Equation (6.6b) then predicts the probe size the microscope is likely to produce when well adjusted, or the probe size can be looked up, for certain I_p / I_c ratios, in Figure 6.1. In practical terms, when aiming for the highest spatial resolution, one is likely to choose I_p ~0.1–0.3 I_c, when aiming for a noise-free analytical signal, one is likely to opt for I_p ~0.5–1.0 I_c.

The performance of a particular microscope optimized in the above way should remain constant from one experimental session to the next, but of course different microscopes can be substantially different from each other.

6.3 STEM IMAGE FORMATION

6.3.1 Basic Principles

STEM imaging has several advantages when compared with its fixed-beam counterpart (usually denoted as CTEM, where the "C" stands for "conventional"), at both low and high primary energies [1, 3, 5].

The first advantage is its great versatility. The STEM forms a small probe with pre-sample optics and can thus use any signal coming off the sample to form images. Many different signals, such as elastic and inelastic scattering as well as signals generated within the sample (X-rays, visible photons, secondary and Auger electrons, beam-induced current, etc.) are potentially available. They provide information about the sample's structure, projected potential, chemical composition, atomic bonding, and optical and electronic properties. Acquiring the different signals simultaneously keeps the radiation damage much smaller than if a separate exposure was performed for each type of image.

Another significant advantage is that in the STEM, electrons that have been scattered to high angles or which have lost more than a few tenths of eV of energy can be used to form images without any loss of resolution or collection efficiency. In the CTEM, by comparison, image formation with such electrons typically results in an unacceptable loss of either resolution or collection efficiency. A further advantage comes from the fact that the STEM probe shape (and ADF image resolution) is determined by the interference of electron wave components that have passed through opposite sides of the illumination aperture, whereas the resolution of the usual CTEM bright field (BF) images is determined by the interference between the unscattered electron beam in the center of the CTEM objective aperture and the scattered waves near the edges of the aperture (real or effective, the effective aperture being determined by contrast damping envelopes). The aperture size is limited to similar angular diameters for both STEM and CTEM by microscope aberrations. However, the larger separation between the interfering wave parts in the STEM case results in STEM ADF resolution that is about 1.6 times better than for CTEM BF imaging in a microscope with the same optical parameters, and a similar improvement for STEM annular bright field (ABF) imaging.

Two other major advantages of STEM ADF imaging come from the fact that the ADF signal is predominantly produced by electrons that have been scattered by the atomic nuclei. This makes the ADF intensity of individual atoms depend on the charge of the atom's nucleus. Different Z atoms therefore give very different ADF image intensities, and it is not difficult to tell which atom is which for well-resolved atoms. In bright field (BF) CTEM, by contrast, images of individual atoms of different Z can have similar intensities, and the chemical type of the atoms is often impossible to distinguish. Moreover, because the atomic nucleus is much smaller than the projected potential that is imaged by BF CTEM, and also smaller than the valence electron distribution imaged by scanning tunneling microscopy (STM) and atomic force microscopy (AFM), the localization of the ADF signal is much better than the localization of the signals used by the other techniques. This is now allowing STEM to reach higher resolution on many types of samples than regular CTEM and also STM and AFM. In the future, when the STEM instrumental resolution improves further and thus catches up even more closely with the excellent localization of the ADF signal, ADF STEM images of isolated heavy atoms are likely to become $\ll$1 Å in size.

In practice, the various signals that can be collected of course depend on what kinds of detectors have been built into the STEM. Figure 6.2 shows three of the many possible detector configurations.

Configuration (a) corresponds to the way most STEMs are run. The high-angle ADF (HAADF) signal, typically spanning 80–200 mrad collection half-angle (at 60–100 keV), is used to locate the atoms and to determine how heavy they are. The medium angle ADF signal (MAADF), typically 40–50 to about 80 mrad can be used to probe elastic strains in crystalline samples. Adding it to the HAADF signal, or changing the projector regime so the inner angle on the "HAADF" detector becomes around 50 mrad, gives stronger signal from light atoms in ultra-thin samples such as graphene. The EELS, typically spanning up to about 50 mrad collection half-angle, is used to determine the chemical composition of the sample, and also its electronic properties, by analyzing the fine structure of EELS edges and also of low-loss spectra. The illumination semi-angle is typically around 30–35 mrad in aberration-corrected STEM, and the EEL spectrometer therefore typically collects all electrons scattered by less than about 15 mrad.

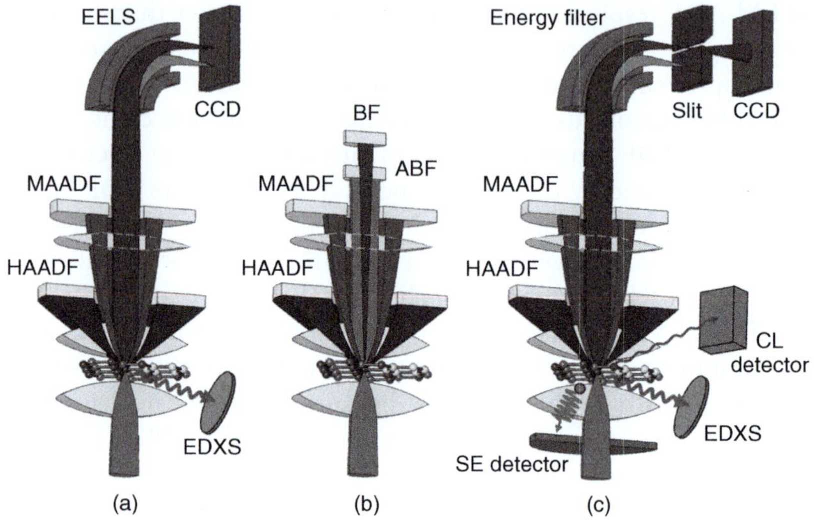

Figure 6.2 Different ways of configuring detectors in a STEM in which the electrons travel upwards. (a) "Standard" materials science set-up combining ADF imaging (either HAADF or MAADF, or both) with EELS and optionally also EDXS. (b) Set-up for imaging light elements using annual bright field (ABF) and on-axis bright field (BF) detectors. (c) The "physicist's set-up" for capturing and analyzing 2D, energy-filtered diffraction patterns plus several other analytical signals (See plate section for coloured version).

The reason this configuration has become "standard" is that it can determine the sample structure, composition and atomic bonding properties all at the same time. In materials science-focused laboratories, it is often enhanced by an energy-dispersive X-ray spectrometer (EDXS), which either supplements or even replaces the EEL spectrometer.

The configuration shown in Figure 6.2(b) is especially useful for imaging light elements. The BF and annular bright field (ABF) signals are sensitive to the sample's projected potential [25, 26], and hence are well suited to detecting very light atoms such as lithium and even hydrogen, especially in a matrix of heavier atoms [27–29]. They are also well suited to exploring charge transfer from one atomic species to another. The ABF detector is to a large extent complementary to the axial BF one, but it has two important advantages: it can give higher spatial resolution, and its collection efficiency is higher. It shares a disadvantage of BF images: the imaging is coherent and varying the defocus and sample thickness gives rise to many different image forms, which means that trustworthy image interpretation is only possible if exhaustive simulations are carried out. Although the ABF detector has so far been mostly used at primary energies of 100–200 keV, using it at lower energies should reduce the knock-on damage and allow images with better statistics to be acquired.

When the probe is focused on the sample, the detection plane contains a two-dimensional diffraction pattern, which changes as the probe advances from pixel to pixel. Recording this complete data set in a "diffraction-image" akin to a spectrum-image [30] is very attractive, but there is a problem: uncompressed, the diffraction-image represents a huge amount of information. Recording the diffraction pattern with a 1024×1024 pixel detector at every pixel in a 1024×1024 image amounts to 4 terabytes (3.5×10^{13} bits) of data if the intensity at each pixel is stored as a real number. Such a data set would be greatly redundant and rather sparse, if acquired in a reasonable time: a 1000 s acquisition with a 100 pA beam current would only deliver 6×10^{11} electrons to the sample, that is, 56 times less than the number of bits assigned for storing the data.

Realistic detection strategies have therefore concentrated on dividing the diffraction plane into segments that separate the various signals in such ways that images formed with them provide information about different properties of the sample. Several such strategies have been put into practice, for example, Cowley *et al.*'s STEM detector system [31] that allowed the two-dimensional diffraction pattern to be recorded by a TV camera and selected parts of it to be sent to fast photomultiplier (PMT) detectors, Haider *et al.*'s segmented solid state detector which used a system of concentric annuli subdivided into four radial segments each [32], and Shibata *et al.*'s system of scintillators coupled by optical fibers to a bank of 16 photo-multiplier tubes (PMTs) that are read out simultaneously [33]. An ability to detect the whole diffraction pattern at every pixel is available if a CCD camera or an imaging filter [34, 35] is mounted on a STEM, and with the second option, the pattern can be energy filtered if needed. The read-out is then of course rather slow: typically about 2–100 patterns per second, which restricts the pixel advance rate even more than EELS spectrum-imaging [36, 37]. But it should be possible to improve this rate in the future with newer and faster camera designs.

The approach is illustrated by Figure 6.2(c). In most cases, only the coherent part of the diffraction pattern, limited to electron momentum transfers smaller than about 2 Å^{-1}, which corresponds to the part of the diffraction pattern spanning up to about 100 mrad at 60 keV (and 50 mrad at 200 kV) has richly detailed information. This means that an HAADF detector, and sometimes even an MAADF detector, can be used at the same time as recording the diffraction pattern, without a loss of key diffraction data. The shown imaging filter is able to select the energy of the electrons producing the two-dimensional diffraction pattern. Other detectors, shown near the sample, enrich the acquired data sets with secondary electron images, cathodoluminescence images and spectra, and elemental maps acquired with an EDXS detector.

Recent work by Kimoto and Ishizuka [38] has shown the power of acquiring and analyzing the diffraction pattern at every pixel with a two-dimensional detector. Among other findings, they were able to demonstrate that electrons detected in different parts of the pattern went past atomic columns in different average locations x,y in the sample plane, and that a small improvement in the resolution of the images of atomic columns would be possible if this effect was taken into account. The work showcased the extreme richness of the information available in the coherent part of the diffraction pattern. The pattern is essentially an in-line hologram, and all the information that is made available by off-axis holography [39] in a CTEM is therefore encoded in it, though it is only recoverable with a greater degree of complication than in the off-axis method.

6.3.2 ADF Imaging

At the other end of the difficulty-of-interpretation scale, when STEM ADF imaging uses only sufficiently high angles, it becomes largely incoherent. Incoherent images have the great advantage that they either show the correct contrast or no contrast at all: the images typically fade out with increasing defocus without changing character. This makes them straightforward to interpret.

The transition from incoherent to partially coherent STEM ADF imaging is illustrated in Figure 6.3. The figure shows two images of the same area of graphene displaying several features: monolayer graphene (at lower left in a and b), double-layer graphene, a domain boundary in one of the layers (indicated by dotted lines in a), some adatoms (arrowed in a and at lower left in b), and a somewhat mobile contamination layer (on the left in b). The graphene was actually prepared from the ^{13}C isotope, that is, the carbon nuclei had an extra neutron compared to regular carbon (^{12}C). This had no effect on image contrast, but it raised the onset energy for knock-on damage to above 95 keV [40].

The inset Fourier transforms show that in the region marked 2, on the left side of the domain boundary, there was a rotation of 3° between

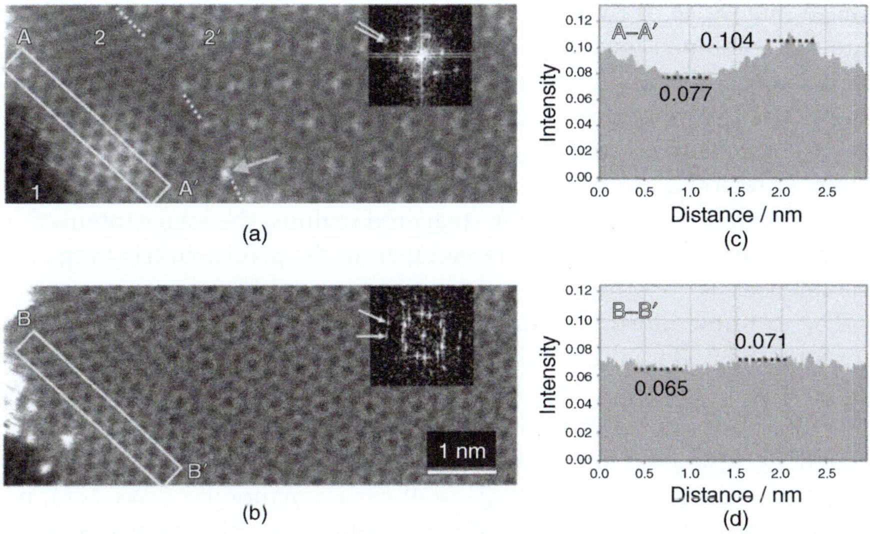

Figure 6.3 Imaging of double-layer graphene at 60 keV with a probe half-angle of 30 mrad and two different signal collection geometries. (a) Partially coherent (37 mrad inner half-angle). (b) Largely incoherent (72 mrad inner ADF half-angle). (c) "Fat" (transversely averaged) profile along the A-A' direction of the long rectangle shown in (a). (d) "Fat" profile along the B-B' direction of the rectangle marked in (b). The numbers in (a) mark the number of graphene layers, the dotted lines mark a domain boundary in one of the two layers, separating the double-layer region 2 from the differently-rotated double layer region 2′. The single green arrow in (a) marks a heavier adatom. A Fourier Transform (FT) taken from the area marked 2 is shown as an inset in (a), an FT from the area marked 2′ is shown as an inset in (b). Yellow arrows pointing to spots in the FTs show maxima due to the two separate layers. Nion UltraSTEM™100, 60 keV. Sample and data courtesy of Professor J.C. Meyer, University of Vienna (See plate section for coloured version).

the two layers, and that in region 2′, on the right side of the boundary, there was a rotation of 19°. As a result, the two parts of the double layer sample region show very different moiré patterns. The patterns are qualitatively similar in (a) and (b), but there is a major difference: some atoms in region 2′ in (a) are imaged with a considerably larger intensity than other atoms, and in region 2 in (a), there is a strong overall modulation of the intensity. As demonstrated by the line profile A-A' shown in (c), the modulation amounts to (0.104–0.077)/0.077 = 35%. A similar profile from the more incoherent image shown in (d) also displays some modulation, but only 9%.

The modulation arises because the imaging is partly coherent, falling between two extreme cases: completely incoherent, in which case the image intensities of two atoms precisely on top of each simply add up and the combined intensity is twice the intensity of a single atom, and completely coherent, in which case the scattering amplitudes for the two atoms need to be added, and the combined intensity is four times the intensity of a single atom. The small rotation between the two graphene layers in region 2 causes the layers to shift gradually from exact overlap to a staggered projection. In the staggered regions, the image intensity is simply twice the intensity of a monolayer; in the precise overlap regions, the intensity is 35% higher.

The example illustrates that STEM ADF imaging is partially coherent longitudinally (in the Z-direction), and incoherent transversely. The longitudinal coherence length is rather short, and the coherent image contribution is not dominant even for the present case of atoms separated by just 3.4 Å in the beam direction (the interlayer distance in graphite). When imaging a single wall carbon nanotube (SWCNT), no coherent imaging effects arising due to projected overlap of atoms in the nanotube's front and back walls, separated by 1.7 nm, were observed with an inner ADF cut-off angle of 55 mrad. (See Figure 15–5 of Ref. [18], which shows a chiral SWCNT imaged at 60 kV and looking rather similar to the image of region 2′ shown here, but with no sign of precisely overlapping atom pairs appearing brighter than other pairs overlapping less precisely.)

The partial coherence makes accurate image interpretation for samples thicker than a monolayer more difficult than if the ADF images were completely incoherent. But even with an inner detector angle only 23% larger than the illumination angle (37 v 30 mrad) and the atoms that give rise to the coherent effects separated by just 3.4 Å in the beam direction, incoherent imaging still remains the dominant term. Moreover, unlike with BF imaging, the partially coherent ADF image does not reverse its contrast when the defocus is changed, which means that intuitive image interpretation is unlikely to lead one to completely wrong structural guesses. In other words, avoiding coherent imaging is of course preferable in ADF STEM, but if the inner angle of the ADF detector is chosen such that the coherent image contribution cannot be neglected, the main effect will be that images of precisely overlapping atoms will be more intense than with purely incoherent imaging. This will need to be accounted for when interpreting partially coherent ADF images quantitatively, but the overall contrast of STEM ADF images is likely to remain closely related to the specimen structure.

6.4 GENTLE STEM APPLICATIONS

6.4.1 Single Atom Imaging

Single atom imaging has recently found an ideal support material: graphene. Graphene is strong, nearly indestructible when the incident beam energy is below the knock-on damage threshold, and its atomic positions are easy to work out even in a noisy image. A wrinkle for studying single atoms on this material is that graphene's surface is rather inert, and nothing attaches to it very firmly. But this does not apply to atoms substitutionally embedded in graphene.

Graphene is a fascinating material in itself, with many unique properties [41–43], such as electron mobility higher than in Si MOSFETs (metal-oxide-semiconductor field effect transistors), the work-horses of present day electronic devices. It has been imaged by several different techniques, including TEM [40, 44–49] and STEM [11, 12, 18, 50–55] in its free-standing form, and by AFM and STM when supported on other materials, for example [56]. The more recent imaging studies have resolved the "holes" in the graphene structure, 2.46 Å apart, and several have used the description "atomic resolution" in their titles. But if "atomic resolution" is taken to mean that all the atoms including the nearest neighbors, spaced just 1.42 Å apart, have been resolved according to the Rayleigh criterion [57], (i.e., with a clear intensity dip between the atoms) then the first such images were obtained in 2009 with a 60 kV STEM and published in 2010 [11].

Graphene is thought to be the top candidate for replacing MOSFETs in integrated circuits [58], first for high speed applications and later perhaps even for general use in microprocessors, memory chips, and so on. An important issue that arises with this material is that in larger sheets, without doping, it is a zero band-gap semiconductor and thus not the best candidate for electronic devices. Doping that makes it semiconducting and hence better suited for device applications can in principle be introduced either as adsorbates on the graphene or substitutionally into the grapheme [59]. Various secondary issues arise with the doping, such as a major reduction of mobility when graphene is doped substitutionally, and that graphene changes its electrical properties when surface contaminants are present. The proposed approaches for overcoming these issues include fashioning graphene into narrow ribbons and doping it along the edges of the ribbons [60], and manufacturing a patchwork of graphene interspaced with patches of hexagonal boron-nitride located within the same two-dimensional sheet [61]. The metal contacts to graphene play an important role, and can modify its electrical properties [52, 62]. Deposited metal atoms can also cause graphene etching, especially in the presence of oxygen [55]. In all these problems, the crucial questions are: "what are the impurity atoms in and next to the active graphene sheet?", "where are they located?", and "what changes in the graphene electronic structure do they cause?"

Perhaps for the first time in the history of materials science, we are now in a position to answer most of these questions, atom-by-atom. Never before have we had such sensitive tools for probing the structure and electronic properties of matter, nor as finely tuned methods of fashioning materials into new devices.

Because of it hexagonal symmetry, imaging graphene has the unusual property that a probe containing spatial frequencies corresponding to distances that are smaller than the graphene nearest-neighbor separation of 1.42 Å is required to resolve the nearest neighbors. This is because resolving the neighbors needs the graphene (11–20) reflection, at $(1.23\,\text{Å})^{-1}$, to be strongly transferred into the image, and transfer of the (20–20) reflection at $(1.07\,\text{Å})^{-1}$ is desirable too. For current STEMs, this does not leave much of a safety margin when operating at 40–60 keV, as can be seen in Figure 6.1, and it makes the task of resolving the atomic structure of graphene at 30 keV rather difficult. Even at 60 keV, the probe size increase due to an I_p that is too large has to be carefully monitored, and the residual higher-order aberration coefficients strictly controlled [63]. A radical change in this situation will probably arrive only with the introduction of practical STEM chromatic aberration correctors, or of efficient monochromators that do not cause any appreciable source-size broadening.

Figure 6.4 shows MAADF images of a single Si atom (as identified by EELS) embedded in monolayer graphene acquired at 60 keV with a 100 pA probe current in an UltraSTEM™100 normally giving a coherent current I_c of 0.3–0.5 nA, a probe convergence semi-angle of 30 mrad and an MAADF inner half-angle of 54 mrad [64].

The microscope aberrations were carefully tuned, and images of the same area of the sample were acquired in two different ways: a single scan at 40 μs per pixel, shown in Figure 6.4(a), and 16 scans at 24 μs per pixel aligned (by cross-correlation) and added together, that is, with 384 μs/pixel total acquisition time, shown in Figure 6.4(c). The images were rotated by about 0.6° and straightened by a linear shear of about 0.5%, to make their atomic rows precisely align with the horizontal and

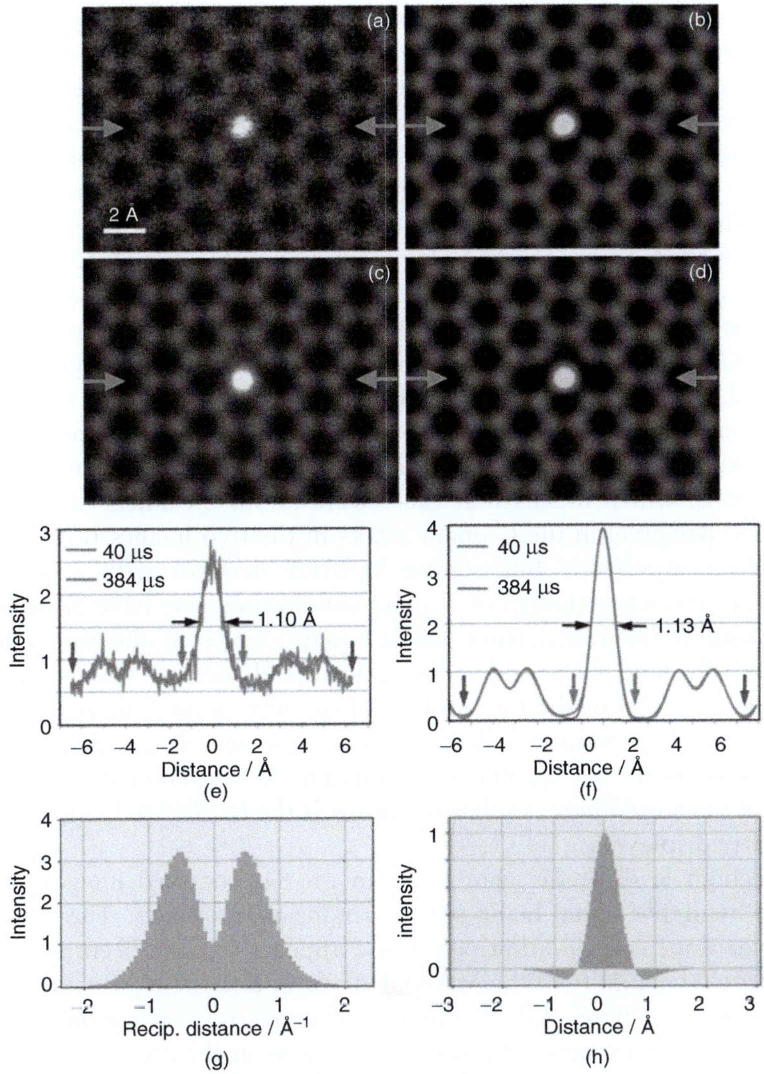

Figure 6.4 MAADF images of a Si atom embedded in graphene taken in two different ways, processed versions of the images, and line profiles through the images. (a) Single MAADF image taken with 40 μs/pixel acquisition time. (b) Image (a) processed with a smoothing and probe skirt-removing filter. (c) Sum of 16 images taken with 24 μs/pixel. (d) Image (c) processed with a smoothing and probe skirt-removing filter. (e) Line profiles taken between the two green arrows in images (a) and (c). (f) Line profiles taken between the two green arrows in images (b) and (d). (g) The smoothing and skirt-subtracting filter as it was used in the FT domain. (h) Real-space shape of the filter. Blue arrows in (e) and (f) mark the centers of graphene rings, red arrows mark ±1.42 Å distance from the center of the Si atom, that is, the C-C nearest neighbor separation. UltraSTEM™ 100, 60 keV (See plate section for coloured version).

vertical directions, and cropped to make them continuously periodic. No other processing was applied to the two images.

The pixel size was 0.039 Å. The exposure level was therefore 1.6×10^7 electrons/Å^2 for the single-shot image (a) and 1.6×10^8 electrons/Å^2 for the summed image. This is about three times and 30 times higher than the dose we used for MAADF imaging of BN [9]. The effective area of each carbon image was about 1.1 Å^2, and hence the above doses are similar to the doses per C atom. The MAADF cross-section for a single C atom [65] imaged at 60 keV with a detector starting at 54 mrad is 1.4×10^{-3} Å^2, and this means that there were about 2.6×10^4 electrons incident on the MAADF detector per C atom in the single-exposure image (a), and about 2.5×10^5 electrons per carbon atom in the summed image (c). However, because of the large oversampling, the electrons detected in the maxima of the images of individual atoms were spread among about 740 pixels, which means that there were about 35 and 340 electrons detected per pixel in the C image peaks in the two images respectively. Assuming a perfectly functioning MAADF detector with a detective quantum efficiency (DQE) of 1, there should therefore have been about 17% and 5.5% r.m.s. (root mean square) noise in the two images respectively (and 85% and 27% peak-to-peak noise). This is roughly what is observed in the line profiles of the unprocessed images, shown in Figure 6.4(e). Note that the profiles are just one pixel wide; the usual practice of using "fat" profiles (integrated in the perpendicular direction) has not been used here, so that the noise at the level of individual pixels is clearly displayed.

The high level of the shot noise in the unprocessed images is due to the available signal being spread among many pixels. The fact that this was done and no additional noise appears to have been picked up testifies to the quality of the MAADF detector and the digitization chain that were used here. A high number of pixels per each atomic image has several advantages [11]: small distortions in the atomic lattice and sudden atomic motions can be readily detected, and one can be sure that there is no resolution loss due to too-coarse a sampling. The eye is good at smoothing the noise out and discerning the overall pattern of the atomic rows, and most STEM images are shown with the shot noise present. But the shot noise does get in the way of quantitative analysis of the acquired data.

Fortunately, the spatial frequencies for most of the shot noise in oversampled images are much higher than the sample's spatial frequencies that have been meaningfully transferred into the STEM image, and the two types of image contributions therefore cleanly separate in the spatial frequency domain. This is the basis for the filtering that resulted in Figures 6.4(b) and (d): the Fourier transforms of the experimental data were multiplied by a rotationally symmetric Gaussian peak corresponding to a real-space full-width at half-maximum of 0.8 Å. This made the filter amplitude >0.3 for the meaningfully transferred spatial frequencies, and close to zero for most of the spatial frequencies of the shot noise.

With the filter applied, the statistical noise in the image is reduced to not much more than the shot noise expected when all the electrons per atom are binned together: 3.1% (peak-to-peak) for the lower-exposure image (b) and 1.0% for the higher exposure image (d). This can be verified by superposing equivalent line profiles from the two smoothed images on top of each other, as is done in Figure 6.4(f). The two images were acquired completely independently, yet the two profiles reproduce each other's atomic maxima within a few percent.

A second problem with the unprocessed images is illustrated by the profiles in Figure 6.4(e): the probe had a "skirt", which spread the atomic signal further away from the center of the atom than a Gaussian-shaped probe would have done. The FWHM of the Si peak is about 1.1 Å in the unsmoothed images. A Gaussian probe of this width would decay to 1.0% of the probe maximum at 1.42 Å away from the center. 1.42 Å being the distance of the center of the six-fold carbon rings in graphene from the ring's atoms, for a Gaussian probe of 1.1 Å FWHM, the intensity in the center would therefore have been about 6% of the C maxima. But this was not the case: the ring centers, marked by the blue arrows in Figure 6.4(e), had about 60% of the intensity of the carbon maxima.

The excess is readily explained by a close examination of the shape of the Si atom peak in the profiles of the unprocessed images. At 1.42 Å away from its maximum, marked by the red arrows, the profile still had about 8% of the maximum intensity (above the "dark level" of about 0.6, as best seen in "fat" profiles of unsmoothed images). Adding six such contributions incoherently, this predicts a 48% intensity in the centers of the hexagon compared to the carbon peaks—a reasonable match. It shows the great sensitivity of graphene imaging to the exact probe shape. It also raises the possibility that the probe shape can be characterized by a careful analysis of graphene images, especially ones that contain non-periodic features.

The excess intensity does not extend very far from the atom. As can be seen in Figure 6.4(e), by about 1.8 Å away from the center of the Si atom, the intensity is essentially down to the same value as the intensity in the centers of the regular graphene holes. The excess "skirt" is partly due to the shape of a diffraction-limited probe produced by a beam that's defined by a sharp aperture: an Airy disc function [66] rather than a Gaussian, and partly due to skirt–increasing modifications of the Airy disc, such as excessive deviations of the aberration function within the aperture. This includes the deviations due to chromatic aberration, which are largely due to the extended tail of the CFEG energy distribution, and which are likely to increase the probe skirt intensity substantially.

The probe skirt causes a problem for the quantification of graphene images at the present level of STEM resolution: the images of individual atoms make non-negligible contributions to the images of their nearest neighbors, which means that the intensity of each atom in the unprocessed image partly depends on what atoms are surrounding it. This was a major issue when trying to identify substitutional atoms in BN by their image intensity [9]: unprocessed MAADF images of BN showed about 20% intensity difference between identical atoms (C) substituting for B and N respectively, simply because neighboring nitrogen atoms contributed more to the image of a C atom at a B site than neighboring boron atoms contributed to a C atom at an N site. The issue, however, had a simple solution [9]: we forced the probe skirt to zero at the nearest-neighbor distance, and the image contribution from the neighbors was thereby nulled. The intensities of the C atoms at the two sites in BN then became similar.

Forcing the probe to zero at the nearest neighbor distance also means that it is forced to zero at the center of the graphene rings, which are the same distance away from the six atoms of each ring as the nearest neighbors are from each other. This gives a convenient metric for judging how much probe skirt to remove: the correct subtraction results in average zero intensity in the centers of the rings. The precise form of the subtracted skirt is not crucially important, and we therefore modeled it by a Gaussian, typically 1.5 to twice as wide as the Gaussian filter used to minimize the image shot noise.

The practical implementation of the skirt-nulling filter is very straightforward: one simply adds a narrower negative Gaussian when multiplying the FT of the image to reduce the shot noise. The profile of the reciprocal space filter used to obtain the smoothed and skirt-subtracted images and profiles (shown in Figure 6.4b, d and f) is shown in Figure 6.4 (g); the profile of the real space version of the filter, obtained by Fourier transforming the reciprocal space filter, is shown in Figure 6.4(h). The filtering procedure is carried out more efficiently in reciprocal space, but its effect is described more easily in real space: it amounts to convolving the experimental image with the real-space function shown in (h). The negative skirt around the central peak in the real-space filter function reaches about −10%, as expected when 60% intensity needs to be removed from the centers of the carbon rings. The efficacy of the double Gaussian filter in "cleaning up" the experimental images is confirmed by the processed images in Figure 6.4(b) and 6.4(d), and the line profiles 6.4(f): the intensity in the center of the graphene rings, including the ones next to the Si atom, is reduced to zero, and the profile of the Si atom begins to resemble a Gaussian peak, with negligible intensity just 1.42 Å away from the center of the atom. The FWHM of the Si atom is increased very slightly by the filtering: from 1.10 Å to 1.13 Å.

The Si impurity atom in Figure 6.4 is substituting for a C–C pair. This is one of two possible configurations for a Si substitutional atom that we have observed graphene, the other one being a Si atom substituting for a single C atom (shown in Reference [14] and in Figures 6.6 and 6.8 later in this chapter). We call the two types of defects type A (Si substituting for one C atom) and type B (Si substituting for two C atoms). This particular type B defect was remarkably stable: it received a dose $>2 \times 10^8$ electrons when the data for Figures 6.4(a) and (c) was recorded, plus an additional amount when the imaging was being set up. Eventually an extra atom was picked up next to the Si atom, thereby transforming the defect into a type A defect [64]. The Si atom then started hopping from one C site to another.

When the contribution of the probe skirt is removed (Figure 6.4f), the Si/C intensity ratio becomes 3.9, which amounts to $(14/6)^{1.61}$. The Si and C atomic intensities I_Z can thus be related to their atomic types Z by a power law of the form:

$$I_z = a\, Z^{1.61} \tag{6.9}$$

where a is a constant.

The value 1.61 is slightly less than the exponent of 1.64 derived experimentally when imaging BN under similar conditions [9]. The difference could easily arise from a small difference in the inner detection angle between the two experiments, or a small error in the subtraction of the probe tail. A part of the difference can also be due to simple statistical variation.

It is interesting to note that the Si/C peak ratio was only about 2.5 in the unprocessed images. The reduced ratio arose because a part of the Si signal was spread over the neighboring C atoms, and the carbon atoms did not make nearly as important a contribution to the Si atom intensity. This can be seen in the uncorrected images, in which the carbon atoms immediately next to the Si atom are brighter than carbon atoms further away.

The discussion above illustrates that with stable samples and large incident doses, ADF image analysis can be very quantitative: the relative intensities of the atoms can be determined to within a few %. The difference in image intensity between a carbon atom and a nitrogen atom is $(7/6)^{1.61} - 1 = 28\%$, and different atomic species within a graphene monolayer are therefore very easy to distinguish and identify in such images. Moreover, with $>10^4$ scattered electrons contributing to an image of a single atom about 1 Å wide, the statistical error in the mean position of the atomic image should be less than 1 pm. In practice, of course, sample shakes and other instabilities will move the atomic images by more than 1 pm in a sequentially scanned image, but such effects can be minimized by recording many scans and summing them up, as was done here. The precision of the determination of the atomic positions from STEM ADF images should then equal or even surpass similar determinations based on CTEM BF imaging, with STEM ADF having the additional advantage that image interpretation is simpler because the imaging is largely incoherent.

Figure 6.5 extends the analysis of the image of Figure 6.4(c) further. Figure 6.5(a) shows that the recorded image can be constrained to be precisely periodic: copies of the rotated and stretched image have been pasted to the right and below the original image, yet the seams, indicated by the yellow arrows, are largely invisible. Note that the stretching was linear, which means that if distortions were present say due to 60 Hz stray fields moving the probe or the scan being non-linear, they would not have been removed. This illustrates progress with the microscope used for the present experiments. The BN data of Reference [9] was acquired using the same microscope, when it was located in a lab with about 0.3 mG (r.m.s.) 60 Hz AC stray fields, and the images had to be re-mapped using a sinusoidal de-warp function to take out the resultant periodic distortion. The present images were taken with the microscope located in a new lab with <0.1 mG stray fields, and no sinusoidal de-warping was necessary.

Figure 6.5(b) shows an FT of the experimental image. Because the image has been made periodic, it has no sharp boundaries. As a result, there are no horizontal or vertical streaks in the FT of the type that are present in FTs of images that do have sharp discontinuities. The absence of vertical streaks also signifies that the probe was exceptionally stable. In STEM image acquisition, the probe often wobbles slightly relative to the sample, and this causes successive scan lines to be randomly

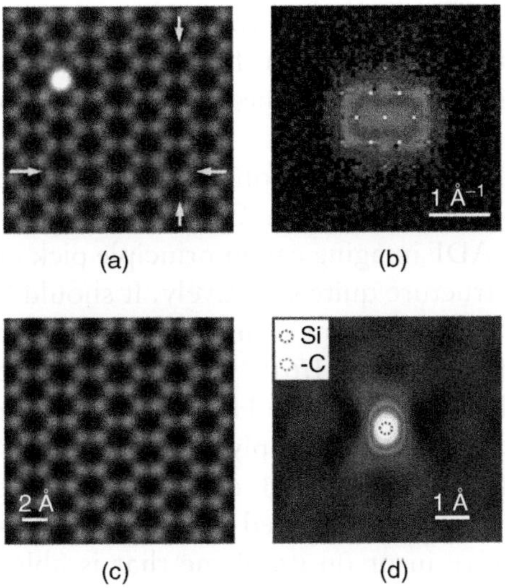

Figure 6.5 Processing of the sum of 16 images shown in Figure 6.4(c). (a) Image that's been stretched and rotated to make it continuously periodic; the yellow arrows mark the imposed periodic boundaries. (b) Central part of the Fourier transform of the periodic image, showing graphene reflections at (2.13 Å)$^{-1}$ (yellow spots), (1.23 Å)$^{-1}$ (orange spots), and (1.07 Å)$^{-1}$ (green spots). (c) Image of "perfect graphene lattice" obtained by masking off the non-periodic component in (b) and performing an inverse FT. (d) Image of the non-periodic component, showing the Si atom at the center, two "minus" carbon atoms, and the non-periodic variations of image intensities at the C sites in the graphene (See plate section for coloured version).

displaced by a few tenths of Å with respect to each other. It results in "jiggles" visible at a high magnification (see for instance Figure 1(b) in Reference [4]): the atomic images have serrated edges rather than smooth ones, because of the random displacement of successive scan lines. Such an image contains sharp discontinuities between neighboring scan lines, and this results in FT streaks running at right angle to the scan direction. No such streaks are present in the present FT, which means that the probe wobbles were most likely <0.1 Å r.m.s., in agreement with our earlier observations [18].

Filtering out the spots due to the periodic lattice results in an image that shows only the non-periodic parts. Such an image, which has additionally been processed with the smoothing and skirt-subtracting filter of Figure 6.4(g), is shown in Figure 6.5(d). The non-periodicities include the Si atom, two "minus" C atoms (the atoms that the Si atom replaced show up as negative atoms in an image of the deviations from periodicity), and the variations in the intensities of the carbon atoms in the regular graphene sheet. The filtered-out periodic component is shown in Figure 6.5(c), in which the Si atom has been replaced by two carbon atoms.

The contrast variations in the non-periodic image away from the Si atom amount to $<1\%$ r.m.s. of the C atom intensities in Figure 6.4(d). This confirms that ADF imaging can in principle pick up small variations from the average structure quite sensitively. It should be able to detect a single stationary hydrogen atom sitting in a non-overlapping configuration, which would have given $(1/6)^{1.61} = 5\%$ of the intensity of a carbon atom, and possibly even detect the H atom when overlapping with the projected image of a C atom, simply by the extra 5% of intensity. A combination of an ADF image with excellent statistics with an ABF or BF image of the same area, acquired simultaneously, should be able to identify any structure in or on graphene that is able to withstand the dose needed for the good statistics.

Examining Figure 6.4(d) carefully shows that the four carbon atoms that are the nearest neighbors of the Si atom have slightly different intensities, causing small deviations from mirror-symmetry of the imaged defect about a horizontal axis. The two C atoms below (in the image) the Si atom are slightly brighter, the two carbon atoms above the Si atom are slightly less bright. Opposing this asymmetry is another one: the five-fold ring with the Si atom at its apex located above the Si atom has a slightly brighter center than the ive-fold ring below the atom. There is also a small left-right asymmetry, with the C atoms just to the left of the Si atom a little brighter than the atoms on the right.

Density functional theory (DFT) calculations did not show any atomic rearrangement that might explain the asymmetry. A combination of stationary or even moving adatoms that might explain it is not at all obvious. However, Figure 6.5(d) provides a likely explanation for its origin. The shown non-periodic component of the image, which amounts to one Si atom plus two negative carbons, is not precisely mirror-symmetric, and the asymmetry, visible as a "blue skirt" of the atom, has a three-fold astigmatism-like character: it resembles a three-pointed "star of Mercedes". The character of the star is modified by the two "minus" carbon atoms next to the Si atom, but even so it is clear that there was a slight excess of intensity at 12 o′clock (straight up from the Si atom) and at 5 and 7 o′clock (to bottom right and bottom left). It corresponds to the pattern roughly expected from mild three-fold astigmatism (either $C_{2,3}$ or $C_{4,3}$) affecting the exact probe shape. The observed alignment of the three-fold astigmatism star had the result that the images of the bottom two of the Si atom's four carbon neighbors received an extra image contribution from the Si atom, but not the images above it. At the same time, the alignment of the star relative to the regular graphene lattice was such that the nearest neighbors of all the regular carbon atoms were located half-way between the star's maxima and minima, and there was therefore no major modification of the image of the regular graphene. A small left-right asymmetry of the Si image is also visible, of a type that is readily explainable by a small amount of coma ($C_{2,1}$).

The amplitude of the three-fold protrusions from the Si atom image needed to account for the asymmetry of the carbon nearest neighbors is small—only about 2% of the Si atom peak intensity. But the effect of even such a weak skirt is clearly visible on the image of the non-periodic component. This raises the hope that artifacts associated with a non-round shape of the probe can be characterized and removed from incoherent ADF images by simple deconvolution. Similar procedures are applied in advanced CTEM image processing, in which the recorded images are used to determine the imaging parameters more accurately than autotuning was able to do, and the parameters are then used to correct images a-posteriori [48]. In other words, aberration-corrected CTEM imaging is routinely made "more perfect" by compensating for microscope tuning that was slightly incorrect. To our knowledge, no such procedure has yet been used for STEM imaging. It would be very useful when trying to make the imaging as perfect as possible, for example, when wanting to determine the positions of substitutional atoms with pm-level precision.

The single Si atom imaged here is of course not unique: any isolated heavier atom in or on a light atom matrix can be imaged and used to draw conclusions about the imaging in general and the probe shape in particular. However, the atom being anchored so firmly that it could withstand a dose of 2×10^8 electrons without moving, and its atomic environment being known so precisely that it could be subtracted away nearly completely, did make this particular atom rather special.

As a final point for the imaging section of this chapter, it is worth noting that at resolutions approaching and surpassing 1 Å, the atoms themselves make a non-negligible contribution to their image width. This was pointed out nearly four decades ago by Beck and Crewe, who showed that a carbon atom imaged in a 100 keV aberration-corrected STEM with a 30 mrad incident probe semi-angle gives an ADF image with a FWHM that is 10% broader than if the atom were a point scatterer [67].

6.4.2 Single Atom Spectroscopy

Atom-by-atom imaging is interesting and informative, and it leads to more questions: what kind of atoms are we imaging, and what are their properties? It is now possible to answer these questions about essentially any type of a single atom, using spectroscopic means. We review these advances next.

The number of inner shell ionization events N_i expected per second from a single atom with an ionization cross-section σ_i is given by:

$$N_i = n\sigma_i/A \tag{6.10}$$

where n is the electron probe current expressed as electrons per second, and A the cross-sectional area of the probe. $n = 1.25 \times 10^9$ e^-/s for an electron probe of 200 pA. Inner shell ionization cross-sections range from about 10^{-5} to 10^{-2} Å^2 for 60 keV electrons [68], with the higher number applicable to low-energy edges (50–200) eV, especially those produced by heavier atoms that have many electrons giving rise to the edge. The above numbers make it clear that a 1 Å^2 probe with 200 pA of current that is centered over a single atom will be producing 10^4–10^7 inner shell ionization events per second per atom. If the atom is able to withstand such intense bombardment without running away, and the microscope has an efficient electron energy loss spectrometer that is able to accept nearly all inelastically scattered electrons, energy loss spectra that have statistics and energy resolution that are good enough for fine structure studies of individual atoms should be readily obtainable.

The count rate for energy-dispersive X-ray spectroscopy (EDXS), which is often used in addition to, or instead of EELS, is reduced by several additional factors

$$N_X = N_i[\omega G\zeta] \tag{6.11}$$

where ω is the fluorescence yield (the probability that an ionized atom will release its excess energy via the emission of an X-ray as opposed to other de-excitation events), G the geometrical collection efficiency of the X-ray detector, and ζ the probability that an X-ray of the relevant energy impinging on the detector will be detected (sometimes also called the detector's DQE). The reduction of the X-ray signal relative to the EELS signal, represented by the term in the square bracket, is typically quite large, ranging from about $1/10^6$ for low-energy X-ray peaks (<300 eV) analyzed by an ultrathin-window detector with a solid angle covering 0.1 sr, to $1/10^2$ for medium energy X-ray peaks (1–10 keV) detected with a better solid angle.

The Z-dependence of the EDXS/EELS ratio (increasing with increasing Z) largely counteracts the Z-dependence of the EELS cross-section (decreasing with increasing energy). This has the welcome result that the X-ray detection rate does not vary greatly across the periodic table, typically by $\sim$10 times across much of the table (with Z = 1–6 (H-C) being major exceptions). For a 200 pA 60 keV beam about 1 Å^2 in size and a 0.1 sr detector with an ultra-thin window, the X-ray count rate is expected to be between about 1–50 detected X-rays per second per atom (C and above). The fact that all the elements give similar count rates makes EDXS analysis especially well-suited for the case of heavy elements that do not have well-defined EELS edges at energy losses <1500 eV, such as third row transition metals, $_{72}$Hf to $_{85}$At. Another advantage of the X-ray technique is that it readily covers the energy region from about 150 eV–20 keV. EELS normally only covers about 2–3% of the primary energy, that is, energy losses up to about 1.2–1.8 keV at 60 keV.

Figure 6.6 demonstrates that the above orders of magnitude are borne out in practice. It shows the results of EELS and EDXS analysis carried out on a single Si atom embedded in a monolayer graphene sheet [15], at 60 keV, with a 190 pA probe of about 1.4 Å in diameter. A background-subtracted EEL spectrum and an EDX spectrum acquired simultaneously are shown in Figure 6.6(a) and (b) respectively. The atom was first identified in a rapidly recorded survey image, the beam was centered over it, and the two spectrum acquisitions were started, quasi-simultaneously. The atom was continuously tracked during the spectrum acquisition, using a 6 × 6 Å scanned sub-area and manual re-centering whenever the atom jumped or drifted away from the center. This reduced the total count rates substantially: a 1.4 Å diameter probe spends only 4% of its time over the atom of interest when it is scanning continuously over a field of 6 × 6 Å. An ADF survey image was collected both before and after the spectrum acquisition. Figure 6.6(c) shows a survey image collected in 2 s, just after the spectrum acquisition.

The tracking window is shown as an insert in Figure 6.6(b). Tracking was necessary because the Si substitutional atoms tended to jump to neighboring graphene sites about once per 10 s of the sub-scan imaging (i.e., about once per 0.4 s of the time the beam spent over the atom). Because of the small size of the jumps (typically to the nearest neighbor site), the tracking procedure managed not to lose the atoms. It also compensated for slow stage drift of the order of a few Å per minute (in a microscope room without tight temperature control).

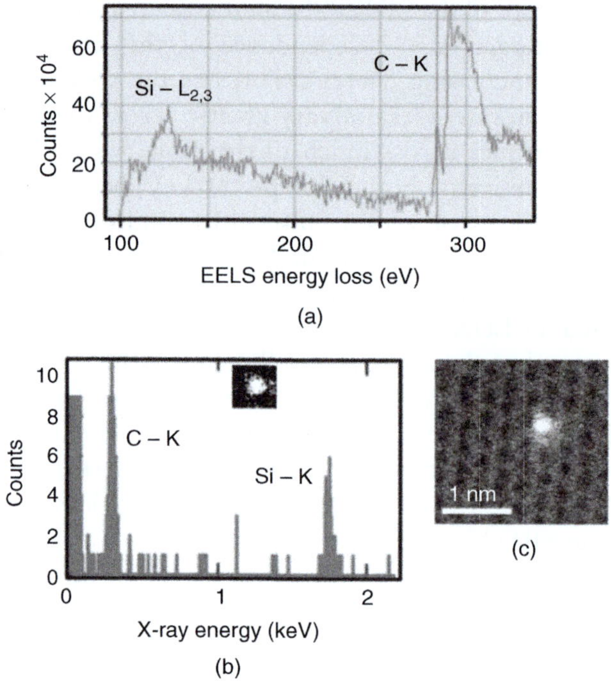

Figure 6.6 EEL and EDX spectra collected from a single Si atom. (a) EEL spectrum collected from a Si atom embedded in monolayer graphene acquired while the atom was being tracked using a 6 × 6 Å tracking window. (b) EDX spectrum collected simultaneously. (c) ADF survey image of the Si atom acquired just after the spectra were collected, with 2 s acquisition time. The small image insert in (b) shows one of the many tracking images acquired in 60 ms each. $I_p = 190$ pA, total acquisition time = 224 s, time the probe spent directly over the atom ~10 s. Nion UltraSTEM™100, Gatan Enfina EELS, Bruker XFlash 5030T EDX spectrometer, 60 keV (See plate section for coloured version).

The total spectrum acquisition time was 224 s, of which about 10 s were spent directly over the Si atom, which therefore received a dose of about 1.1×10^{10} electrons. The EEL spectrum (Figure 6.6a) shows nearly 4×10^5 counts per 0.7 wide energy channel near the maximum of the background-subtracted $L_{2,3}$ edge from the single Si atom, and a total of 3.2×10^7 counts in an energy interval spanning 100 eV from the edge threshold at 99 eV. For the carbon K-edge, the first 100 eV above the edge threshold contained 4.0×10^7 counts. The electron conversion efficiency was 5.0 counts per primary electron, which means that the number of electrons in each energy channel at the silicon maximum detected per each second the beam was over the atom was about 8×10^3, and the integrated silicon edge detection rate amounted to about 6×10^5 detected $L_{2,3}$ electrons per second. The integrated $L_{2,3}$ rate for the single Si atom is about 60% of the total net Si $L_{2,3}$ edge count that would be recorded with the edge integration window extended to infinity, and with 100% efficient EELS signal collection [15]. The actual measured rate is in the upper half of the estimated range of total count rates, as one would expect for an edge at 100 eV that is produced by a total of six L-shell electrons.

When estimating the maximum count rates, we need to consider that an atom can be ionized by an electron passing some distance away from it (delocalization). This effect reduces the peak count rate obtainable with the probe directly over individual atoms, very much as if the probe itself was broadened to match the delocalization diameter. The effect is especially important for low energy ionization events, such as the Si $L_{2,3}$ edge and C-K EELS edge (and X-ray peak) recorded here. For ionization events of 1 keV and greater energy and electron probe sizes of about 1 Å, the delocalization is typically less than the size of the probe [18].

It needs to be noted that the EEL spectrum was read out every second during the acquisition, and that when summing the partial spectra, an automatic dark current subtraction routine introduced noise considerably in excess of the expected shot noise. In other words, the EELS noise level of Figure 6.6(a) is not fundamental and it could have been avoided with better signal processing. Had this been done, the statistical shot noise in the subtracted spectrum should have been about 0.5% r.m.s., that is much less than the about 5% r.m.s. noise in the shown spectrum. However, even with the increased noise level, the form of the Si edge is clearly identifiable. As we will see later on in this chapter, the form is sensitive to the local atomic environment of the single Si atom.

The EDX spectrum (Figure 6.6b) shows clear carbon K and Si K peaks at 0.277 keV and 1.74 keV, respectively. 51 Si counts were recorded in the Si peak, and 115 counts in the C peak. Because of the tracking, the probe spent about 10 s over the Si atom and about 210 s over the carbon region of the tracking window, which contained about 10 carbon atoms. This means that the net count rate was five X-rays per second from the single Si atom, and about 0.5 counts per second per carbon atom. In the same energy windows, in a 256 second scan of pure graphene, 146 carbon counts were recorded, and two counts in the silicon energy window [15]. For either spectrum, ~20 copper-K counts (~8 keV) were also recorded, most probably due to scattered electrons exciting X-rays from the Cu grid the graphene film was supported on. No other peaks were apparent over the entire 0–20 keV detection window.

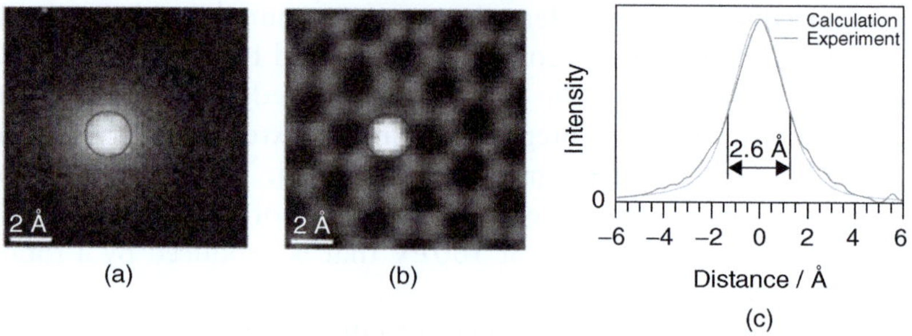

Figure 6.7 Measurement of the point spread function (PSF) for the Si-L edge signal obtained from a substitutional Si atom in monolayer graphene. (a) Si-L edge map computed using an energy window of 100–140 eV. (b) ADF image collected at the same time as the spectrum-image. (c) comparison of the profile of the experimentally measured Si-L map with a theoretical model. Nion UltraSTEM™100, Gatan Enfina EELS, 60 keV (See plate section for coloured version).

The X-ray detector used for this work was not optimized[15]: it had an ultra-thin window that reduced the C count by about 70%, its SDD detector was 30 mm^2 in size, and it subtended only about a 0.1 sr solid angle. But with a stable aberration-corrected STEM column delivering 190 pA probe current into a 1.4 Å probe at 60 keV, the detector readily identified isolated single Si atoms. It also easily identified single Pt atoms[15], and it is very likely that it will be able to identify isolated and stable single atoms of most other elements in the periodic table in the future. These results show that single atom spectroscopy is now a reality for both EELS and EDXS, and that the EDX spectroscopy is going to become even more competitive in the future.

The delocalization of the inelastic scattering, that is, of the primary ionization event, is explored in Figure 6.7. Figure 6.7(a) shows an EELS map extracted from a spectrum-image data set acquired from a single Si atom embedded in monolayer graphene [64]. The spectrum image acquisition parameters were: 60 keV, probe half-angle 30 mrad, 100 pA probe current in a probe about 1.2 Å wide, pixel size 0.25 Å, 0.05 s per-pixel dwell time, spectrometer acceptance half-angle ~50 mrad, EEL spectra collected with 1340 energy channels each 0.3 eV wide. The Si map was extracted from the spectrum-image data set using an energy window spanning from 100–140 eV, and the result was filtered using principal component analysis (PCA) with seven principal components.

Figure 6.7(b) shows an ADF image acquired at the same time as the EELS spectrum-image. It shows the C atoms clearly resolved, which confirms that the probe size was <1.4 Å. The Si atom is also clearly resolved and can be seen to occupy a graphene site in which it displaces two carbon atoms and has four nearest neighbor carbon atoms, that is, type B defect according to our classification. The graphene lattice is mildly distorted, due to the scan being much slower than it was for the images of Figure 6.4, and a small amount of sample drift.

Figure 6.7(c) shows a line profile through the Si map, compared to a theoretical point spread function for the Si L-edge based on a quantum mechanical simulation [64]. The experimental FWHM of the Si image peak is 2.6 Å, which is in excellent agreement with the theoretical calculation. The delocalization measured directly from this experiment is much smaller than expected from simple theoretical arguments [69]. It gives rise to a hope that atomic resolution elemental mapping (with better than 2 Å spatial resolution) may be possible for inner shell losses greater than about 200 eV.

Two further details in the data shown in Figure 6.7 are worth remarking on. A first interesting detail is that the Si atom was directly under the electron beam for about 0.9 s. This means that that the mapped Si atom received a dose of 6×10^8 electrons, that is, only about 5% of the dose received by the Si atom of Figure 6.6, yet it gave a clear spectroscopic image. A second interesting detail is that the Si atom's profile in the elemental map is about two times wider than the probe size. Adding the probe size and the delocalization in quadrature means that the delocalization by itself was at about 2.3 Å FWHM, that is, that the resolution of the Si $L_{2,3}$ map was mostly determined by the delocalization rather than the probe size. Moreover, the probe was close enough to the Si atom to ionize it for about 4.2 s, not just for the 0.9 s it was directly over the atom. The maximum of the mapped signal was reduced correspondingly, that is, the recorded contrast in the Si map was about five times less than it would have been if the map resolution were only determined by the probe size.

The spreading of the available EELS signal due to delocalization has important implications for EELS mapping of single atoms using low energy edges: their visibility will be diminished due to the delocalization of the map. But it also gives rise to an interesting possibility for reducing radiation damage: with good sample stability and sufficiently sophisticated scan control software [70], one can imagine that the optimum way to analyze an atom with a low-lying EELS edge will be to park the probe a small distance away from it for the EELS acquisition, or to circle around the atom without hitting it directly, while checking quickly, from time to time, that the atom has not moved by more than a fraction of an Å. In this way, the ionization of the atom will proceed much as if the probe was parked directly over it, but the likelihood that the atom will be ejected from the lattice by knock-on scattering, which requires an incident electron hitting the atomic nucleus head-on, should be greatly reduced. Such a technique would be related to aloof low loss EELS described by de Abajo and Howie [71], but instead of the beam standing clear of the sample, it would only stand just clear of individual impurity atoms.

6.4.3 Single Atom Fine Structure EELS

Figure 6.8 compares the fine structure of Si $L_{2,3}$ edges from the type A and type B substitutional silicon defects in graphene that were discussed earlier. Figure 6.8(a) shows an ADF image of a type A defect, Figure 6.8(c) shows an ADF image of the type B defect. The imaged type A defect is actually a minor variation on the standard type A: one of the three six-fold rings at whose intersection the Si atom sits in a type A defect is actually a seven-fold ring. However, since this change involves only the third-nearest neighbors of the Si atom, it is not expected to influence the EELS fine structure greatly. The imaged type B defect is also a variant of the standard type B defect: one of the four carbon atoms that are the nearest neighbors of the Si has been replaced by a nitrogen atom.

The brighter squares near the edges of Figure 6.8(a) are an artifact of the used sub-pixel scanning, during which large pixels corresponding to the size of the squares were scanned at the sub-pixel level before advancing to the next large pixel. The acquisition parameters were [72]: 60 keV, illumination half-angle 30 mrad, EELS collection half-angle 48 mrad, 220 pA beam current. For Figure 6.8(a), the pixel size was 1.25 Å and the beam spent 0.18 s at each pixel, during which sub-pixel scanning was used. For Figure 6.8(c), the pixel size was 0.98 Å, the beam spent 0.1 s per pixel, and sub-pixel scanning was not used.

Figure 6.8(b) shows a background-stripped Si $L_{2,3}$ edge extracted from an area of 2x3 major pixels in the spectrum-image acquired simultaneously with the ADF image shown in (a). Figure 6.8(d) shows a Si $L_{2,3}$ edge extracted from a 2×2 pixel area of the spectrum-image acquired just before the ADF image shown in (c). This means that the doses used for producing the spectra shown in Figure 6.8(b) and (d) were 1.4×10^9 and 5.5×10^8 electrons respectively.

The two spectra are compared to reference spectra obtained from other samples. The fine structure of the Si edge from the type A defect is essentially identical (once account is taken of the different energy scales) to the spectrum from the same type of defect shown in Figure 6.6(a). The edge starts at about 100 eV, there is a weak threshold peak at about

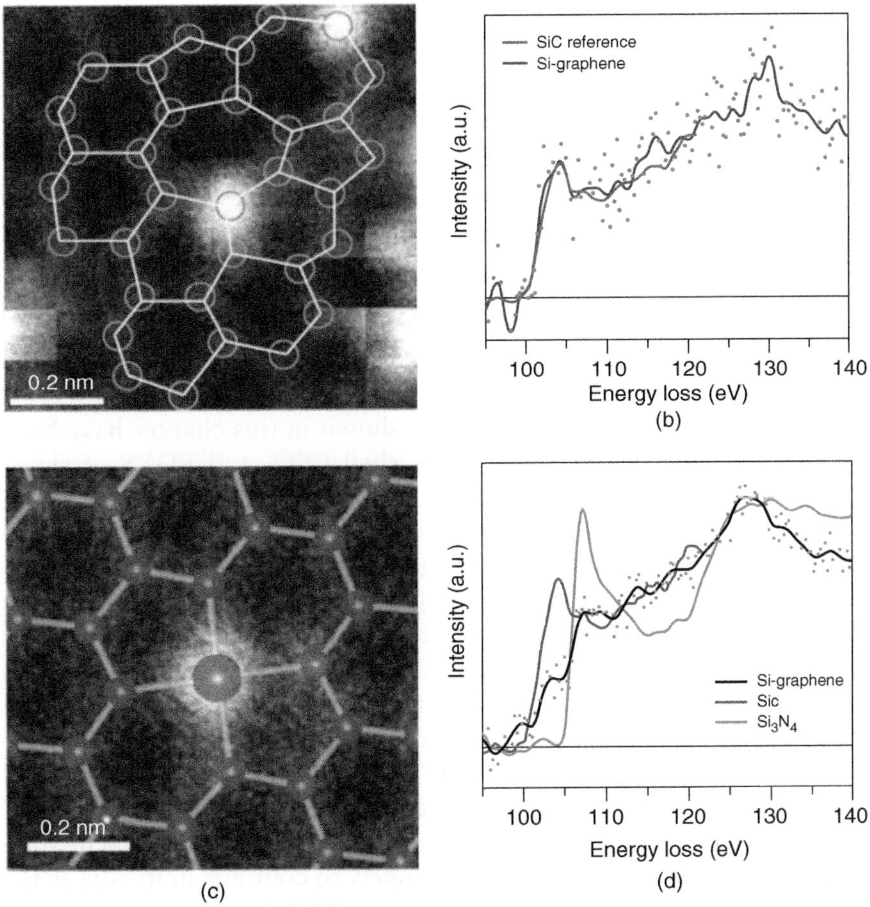

Figure 6.8 Si L-edge fine structures obtained from single substitutional Si atoms in different point defect configurations in monolayer graphene. (a) ADF image of a Si atom substituting for a single C atom (in center of image), in modified type A configuration. (b) Si-L edge from the Si atom shown in figure (a) compared to Si-L edge from SiC. (c) ADF image of a Si atom substituting for two C atoms, in modified type B configuration. (d) Si-L edge from the Si atom shown in (c), compared to Si L-edges from SiC and β-Si_3N_4. Nion UltraSTEM™100, Gatan Enfina EELS, 60 keV (See plate section for coloured version).

104 eV, and a main maximum at about 130 eV followed by an abrupt drop-off. The edge is also very similar to a SiC spectrum shown for comparison in Figure 6.8(b).

The edge from the type B defect is quite different: the threshold peak is absent, and the drop-off past the main maximum is less steep. It is also very different from the SiC and Si_3N_4 edges shown for comparison. DFT calculations are able to model the edge shapes fairly accurately [72].

The ability to probe the fine structure of EEL spectra of individual atoms is likely to find many uses in materials science. The exact edge energy is a useful measure of charge transfer, and the detailed edge shape provides a sensitive measure of the accuracy of theoretical simulations such as DFT modeling that give detailed insights into how the individual atoms are bonded to their neighbors.

6.5 DISCUSSION

For reasons of space, the applications shown in this chapter have been restricted to ADF imaging and inner shell EELS and EDXS of single atoms. Other recent exciting results obtained with gentle STEM have not been described. These include an enhancement of the intensity of the graphene plasmon centered on a single substitutional Si atom [73], and the observation of a one-dimensional EELS resonance at an energy loss of 11 eV at the edges of graphene samples, which appears to amount to a 1D "edge" plasmon that arises in 2D graphene by a mechanism similar to how 2D surface and interface plasmons arise in 3D systems [74]. They provide further evidence of the richness of information that can be expected when a material is probed by an intense, atom-sized beam of electrons in the absence of radiation damage.

Looking towards the future, ADF imaging, at times supplemented by annular bright field imaging (ABF), is likely to continue in its role as the primary structure-probing tool in gentle STEM. Its resolution is likely to keep on improving so that sub-Å imaging will become routine even at operating energies <100 keV.

Of the two principal available atomic-resolution spectroscopy techniques, EELS will continue to give much better count rates and is therefore likely to be the more powerful approach for studies of extremely small amounts of matter. It is likely to be extended, using minimum-dose techniques, to atoms that do not stay completely still under the beam. It may even be possible to perform EELS studies on individual atoms that do suffer radiation damage, provided that the EELS signal is averaged over many different single atoms occupying equivalent sites. Nevertheless, radiation damage is certain to remain a major barrier, preventing many types of samples from being examined as "intensely" as in the examples shown here.

Fine structure EELS studies on individual atoms will be possible at electron doses of the order of 10^9 electrons/Å^2, and they will be especially informative if carried out with doses of 10^{10} electrons/Å^2 and higher. Such doses may be realistic for many types of single atoms, if tracking schemes similar to the one used here (for Figure 6.6) are employed, and the atoms typically only jump to nearest neighbor atom sites while the beam is on them.

The spatial resolution of the analytical techniques will depend on the delocalization of the inelastic scattering, and for larger energy losses, also on the electron probe size. The delocalization depends inversely on the magnitude of the energy loss [18, 69], roughly as $\Delta E^{-1/4}$. The width of the Si $L_{2,3}$ image shown in Figure 6.7 suggests that spatial frequencies of the order of $(2\,\text{Å})^{-1}$ will be transferred into elemental maps resulting from energy losses $\Delta E > 200\,\text{eV}$. A convenient test of the resolution in an elemental map obtained with low energy losses is whether the holes in graphene, 2.46 Å apart, are resolved in a carbon elemental map. This test has now been passed: Figure S9 in Reference [73] shows the holes faintly resolved.

EDXS cannot match the EELS count rates, which means that single atoms are likely to be detectable by EELS using lower doses than those required by EDXS. It also cannot match the EELS energy resolution, but even so it does have important advantages. Because weakly present features stand out more readily above the small EDXS background, the minimum mass fraction (MMF) detectable by EDXS is likely to be better than the EELS MMF, particularly for elements without especially favorable EELS edges. This is likely to make EDXS well suited to detecting single atoms in samples thicker than a few monolayers. Unlike EELS, EDXS will not have any "blind spots" in certain parts of the periodic table (except for hydrogen to lithium), and will therefore be especially useful for detecting elements such as heavy metals that often play crucial roles in catalysis. With similar electron beam characteristics as were used here and an optimized detector design (which we are now developing), EDXS count rates from single atoms are expected to improve by 5–10 times.

High and clean vacuum will be a must-have requirement for this type of analysis, in which a single atom of carbon contamination, or a single adsorbed water molecule that dissociates under the beam and etches the sample, would have altered the results. Further distinguishing characteristics of electron microscopes able to do this kind of work will be an excellent overall stability and a small and energetically narrow electron beam produced by an ultra-bright electron source, combined with high-quality aberration-corrected optics.

In the future, other analytical techniques, some of which are illustrated in Figure 6.2, are likely to join ADF, EELS and EDXS as analytical tools used in many laboratories. Some of them, such as imaging using SE and Auger electrons are likely to match the spatial resolution of EELS and EDXS, while others such as CL and ultra-high energy resolution energy loss spectroscopy are likely to sacrifice spatial resolution for the sake of superior energy resolution.

Another present trend that is likely to become even more pronounced in the future is constructing microlaboratories inside the electron microscope, to perform atomic deposition, etching, direct measurement of electrical properties using STM-type probes, studies of sample-gas interaction, etc., both at room temperature and with cooled or heated samples. Not only are we likely to be building microscopes that can see atom-sized objects more and more clearly, we are also likely to be providing more and more interesting objects for them to examine.

6.6 CONCLUSION

In their classic paper on STEM detection limits [75], published in 1975, Isaacson and Johnson wrote: "So the feasibility of elemental analysis of single light atoms remains a distinct possibility (in theory, at least!)." But attaining such performance in practice seemed a very long way off back then.

Thirty-seven years later, we are not only detecting single atoms, but also studying their bonding by analyzing the fine structures of their EELS edges. We are also analyzing the chemical type of single atoms by X-ray spectroscopy, which Isaacson and Johnson and others showed to be less sensitive that the EELS technique. This remarkable progress is due to several key developments, briefly described in the discussion section above.

Further progress is practically certain, but it is likely to be more incremental: new types of single atoms will be detected and analyzed, image quality will become better, the signal-to-noise ratio and the energy resolution of experimental spectra will be improved, and it will become possible to study single atoms at lower keV, thus further reducing knock-on radiation damage. It seems probable that the early 2010s will be remembered as the watershed years, during which electron beam analysis at the ultimate limit—single atoms—finally became possible, and in many instances nearly routine.

ACKNOWLEDGEMENTS

We are grateful to our colleagues at Nion for designing and building many important parts of the instruments the present work was done with, to Dr R. Zan for the provision of the graphene sample used for the EDXS analysis, to Professor J.C. Meyer for the provision of the sample and data used in Figure 6.3, and to Professor S.J. Pennycook for discussions and provision of laboratory facilities. Research at Oak Ridge National Laboratory was supported by the Materials Sciences and Engineering Division of the Office of Basic Energy Sciences, U.S. Department of Energy (MFC), and Oak Ridge National Laboratory's Shared Research Equipment (ShaRE) User Facility (JCI), which is sponsored by the Office of Basic Energy Sciences, U.S. Department of Energy. Research at Vanderbilt University (WZ) was supported by U.S. National Science Foundation through grant No. DMR-0938330. Research at the SuperSTEM laboratory (QMR) was funded by the UK Engineering and Physical Sciences Research Council (EPSRC).

REFERENCES

1. A.V. Crewe, A New Kind of Scanning Microscope, *J. Microscopie*, **2**, 369–371 (1963).
2. A.V. Crewe, J. Wall and J. Langmore, Visibility of Single Atoms, *Science*, **168**, 1338–1340 (1970).
3. A.V. Crewe, J.P. Langmore and M.S. Isaacson, Resolution and Contrast in the Scanning Transmission Electron Microscope, in: *Physical Aspects of Electron Microscopy and Microbeam Analysis*, eds. B.M. Siegel and D.R. Beaman, John Wiley & Sons, Inc., New York, NY, 47–62 (1975).
4. A.V. Crewe, Direct imaging of single atoms and molecules using the STEM. *Proceedings 47th Nobel Symposium*, *Chemica Scripta*, **14**, 17–20 (1979).
5. S.J. Pennycook and P.D. Nellist, (*eds*), *Scanning Transmission Electron Microscopy -Imaging and Analysis*, Springer (2011).
6. T. Sasaki, H. Sawada, F. Hosokawa, *et al.*, Performance of low-voltage STEM/TEM with delta corrector and cold field emission gun, *J. Electron Microscopy*, **59** (Supplement), S7–S13 (2010).
7. N. Dellby, N.J. Bacon, P. Hrncirik, *et al.*, Dedicated STEM for 200 to 40 keV operation, *Europ. Phys. J. Appl. Phys.*, **54**, 33505 (2011).
8. O.L. Krivanek, G.J. Corbin, N. Dellby, *et al.*, An electron microscope for the aberration-corrected era, *Ultramicroscopy*, **108**, 179–195 (2008).
9. O.L. Krivanek, M.F. Chisholm, V. Nicolosi, *et al.*, Atom-by-atom structural and chemical analysis by annular dark field electron microscopy, *Nature*, **464**, 571–574 (2010).
10. K. Suenaga, M. Tence, C. Mory, *et al.*, Element-Selective Single Atom Imaging, *Science*, **290**, 2280–2282 (2000).

11. O.L. Krivanek, N. Dellby, M.F. Murfitt, *et al.*, Gentle STEM: ADF imaging and EELS at low primary energies, *Ultramicroscopy*, **110**, 935–945 (2010).
12. K. Suenaga and M. Koshino, Atom-by-atom spectroscopy at graphene edge, *Nature*, **468**, 1088–1090 (2010).
13. K. Suenaga, Y. Iizumi and T. Okazaki, Single atom spectroscopy with reduced delocalization effect using a 30 kV-STEM, *Eur. Phys. J. Appl. Phys.*, **54**, 33508 (2011).
14. O.L. Krivanek, M.F. Chisholm, M.F. Murfitt and N. Dellby, Scanning transmission electron microscopy: Albert Crewe's vision and beyond, submitted to *Ultramicroscopy* (2012).
15. T.C. Lovejoy, Q.M. Ramasse, M. Falke, *et al.*, Single Atom Identification with Energy Dispersive X-ray Spectroscopy, *Appl. Phys. Letts*, **100**, 154101 (2012).
16. L. Reimer, *Image Formation in Low-Voltage Scanning Electron Microscopy* (SPIE Tutorial Text Vol. TT12, Tutorial Texts in Optical Engineering, SPIE Press, 1993) ISBN: 0819412066.
17. Y. Zhu, H. Inada, K. Nakamura and J. Wall, Imaging single atoms using secondary electrons with an aberration-corrected electron microscope, *Nature Materials*, **8**, 808–812 (2009).
18. O.L. Krivanek, M.F. Chisholm, N. Dellby and M.F. Murfitt, Atomic-resolution STEM at low primary energies, in: *Scanning Transmission Electron Microscopy: Imaging and Analysis*, S.J. Pennycook and P.D. Nellist (ed.), Springer, New York, NY (2011).
19. J.A. Lin and J.M. Cowley, Calibration of the operating parameters of an HB5 STEM instrument, *Ultramicroscopy*, **19**, 31–42 (1986).
20. O.L. Krivanek, N. Dellby and M.F. Murfitt, Aberration Correction in Electron Microscopy, in: *Handbook of Charged Particle Optics*, 2nd edn, J. Orloff (ed.), CRC Press, Boca Raton, FL (2009).
21. A.R. Lupini, The Electron Ronchigram, in: *Scanning Transmission Electron Microscopy: Imaging and Analysis*, S.J. Pennycook and P.D. Nellist (ed.), Springer, New York, NY (2011).
22. P.E. Batson, N. Dellby and O.L. Krivanek, Sub-Ångstrom resolution using aberration-corrected electron optics, *Nature*, **418**, 617–620 (2002).
23. J.M. LeBeau, S.D. Findlay, L.J. Allen and S. Stemmer, Quantitative Atomic Resolution Scanning Transmission Electron Microscopy, *Phys. Rev Lett.*, **100**, 206101 (2008).
24. C. Dwyer, R. Erni and J. Etheridge, Method to measure spatial coherence of subangstrom electron beams, *Appl. Phys. Letts.*, **93**, 021115 (2008).
25. H. Rose, Phase Contrast in Scanning Transmission Electron Microscopy, *Optik*, **39**, 416–436 (1974).
26. J.M. Cowley, M.S. Hansen and S.-Y. Wang, Imaging modes with an annular detector in STEM, *Ultramicroscopy*, **58**, 18–24 (1995).
27. S.D. Findlay, N. Shibata, H. Sawada, *et al.*, Robust atomic resolution imaging of light elements using scanning transmission electron microscopy, *Appl. Phys. Letts.*, **95**, 191913 (2009).
28. Y. Oshima, H. Sawada, F. Hosokawa, *et al.*, Direct imaging of lithium atoms in LiV_2O_4 by spherical aberration-corrected electron microscopy, *J. Electron Micr.*, **59**, 457–461 (2010).

29. R. Ishikawa, E. Okunishi, H. Sawada, *et al.*, Direct imaging of hydrogen-atom columns in a crystal by annular bright-field electron microscopy, *Nature Materials*, **10**, 278–281 (2011).
30. C. Jeanguillaume and C. Colliex, Spectrum-image: The next step in EELS digital acquisition and processing, *Ultramicroscopy*, **28**, 252–257 (1989).
31. J.M. Cowley and J.C.H. Spence, Innovative Imaging and Microdiffraction in STEM, *Ultramicroscopy*, **3**, 433–438 (1979).
32. M. Haider, A. Epstein, P. Jarron and C. Boulin, A versatile, software configurable multichannel STEM detector for angle-resolved imaging, *Ultramicroscopy*, **54**, 41–59 (1994).
33. N. Shibata, Y. Kohno, S.D. Findlay, *et al.*, New area detector for atomic-resolution scanning transmission electron microscopy, *J. Electr. Microsc.*, **59**, 473–479 (2010).
34. O.L. Krivanek, A.J. Gubbens, N. Dellby and C.E. Meyer, Design and first applications of a post-column imaging filter, *Microsc. Microanal. Microstruct.*, **3**, 187–199 (1992).
35. A.J. Gubbens, M. Barfels, C. Trevor, *et al.*, The GIF Quantum, a next generation post-column imaging energy filter, *Ultramicroscopy*, **110**, 962–970 (2010).
36. M. Bosman, V.J. Keast, J.L. Garcia-Munoz, *et al.*, Two-Dimensional Mapping of Chemical Information at Atomic Resolution, *Phys. Rev. Lett.*, **99**, 086102 (2007).
37. D.A. Muller, L. Fitting Kourkoutis, M.F. Murfitt, *et al.*, Atomic-Scale Chemical Imaging of Composition and Bonding by Aberration-Corrected Microscopy, *Science*, **319**, 1073–1076 (2008).
38. K. Kimoto and K. Ishizuka, Spatially resolved diffractometry with atomic-column resolution, *Ultramicroscopy*, **111**, 1111–1116 (2011).
39. H. Lichte, P. Formanek, A. Lenk, *et al.*, Electron Holography: Applications to Materials Questions, *Annual Rev. Mater. Res.*, **37**, 539–588 (2007).
40. J.C. Meyer, F. Eder, S. Kurasch, *et al.*, An accurate measurement of electron beam induced displacement cross sections for single-layer graphene, *Phys. Rev. Letts*, **108**, 196102 (2012) and http://arxiv.org/abs/1203.2372.
41. K.S. Novoselov, A.K. Geim, S.V. Morozov, *et al.*, Electric Field Effect in Atomically Thin Carbon Films, *Science*, **306**, 666–669 (2004).
42. A.K. Geim and K.S. Novoselov, The rise of graphene, *Nature Materials*, **6**, 183–191 (2007).
43. A.K. Geim, Graphene: Status and Prospects, *Science*, **324**, 1530–1534 (2009).
44. S. Iijima. Observation of single and clusters of atoms in bright field electron microscopy, *Optik*, **48**, 193–213 (1977).
45. J.C. Meyer, A.K. Geim, M.I. Katsnelson, *et al.*, The structure of suspended graphene sheets, *Nature*, **446**, 60–63 (2007).
46. J.C. Meyer, Ç.Ö. Girit, M.F. Crommie and A. Zettl, Imaging and dynamics of light atoms and molecules on graphene, *Nature*, **454**, 319–322 (2008).
47. Ç.Ö. Girit, J.C. Meyer, R. Erni*et al.*, Graphene at the Edge: Stability and Dynamics, *Science*, **323**, 1705–1708 (2009).
48. J.R. Jinschek, E. Yucelena, H.A. Calderon and B. Freitag, Quantitative atomic 3-D imaging of single/double sheet graphene structure, *Carbon*, **49**, 556–562 (2011).

49. J.C. Meyer, S. Kurasch, H.J. Park, *et al.*, Experimental analysis of charge redistribution due to chemical bonding by high-resolution transmission electron microscopy, *Nature Materials*, **10**, 209–215 (2011).
50. M. Isaacson, M. Ohtsuki and M. Utlaut, Can we determine the structure of thin amorphous carbon films using scanning transmission electron microscopy? *Proc. 37th EMSA Meet.*, 498–501 (1979).
51. M.H. Gass, U. Bangert, A.L. Bleloch, *et al.*, Free-Standing Graphene at Atomic Resolution, *Nature Nanotech.*, **3**, 676–681 (2008).
52. R. Zan, U. Bangert, Q. Ramasse and K.S. Novoselov, Metal–Graphene Interaction Studied via Atomic Resolution Scanning Transmission Electron Microscopy, *Nano Letts.*, **11**, 1087–1092 (2011).
53. R. Zan, U. Bangert, Q. Ramasse and K.S. Novoselov, Evolution of Gold Nanostructures on Graphene, *Small*, **7**, 2868–2872 (2011).
54. P.Y. Huang, C.S. Ruiz-Vargas, A.M. van der Zande, *et al.*, Grains and grain boundaries in single-layer graphene atomic patchwork quilts, *Nature*, **469**, 389–392 (2011).
55. Q.M. Ramasse, R. Zan, U. Bangert, *et al.*, Direct experimental evidence of metal-mediated graphene etching, ACS Nano, **6**, 4063–4071. DOI: 10.1021/nn300452y (2012).
56. E. Stolyarova, K.T. Rim, S. Ryu, *et al.*, High-resolution scanning tunneling microscopy imaging of mesoscopic graphene sheets on an insulating surface, *Proc. Nat. Acad. Sci.*, **104**, 9209–9212 (2007).
57. Lord Rayleigh, Investigation in Optics, with Special Reference to the Spectroscope, *Philos. Mag.*, **8**, 261–274, 403–411, 477–486 (1879).
58. F. Schwierz, Graphene transistors, *Nature Nanotech*, **5**, 487–496 (2010).
59. B. Guo, L. Fang, B. Zhang and J. R. Gong, Graphene Doping: A Review, *Insciences J.*, **1**, 80–89 (2011).
60. K. Brenner, Y. Yang and R. Murali, Edge doping of graphene sheets, *Carbon*, **50**, 637–645 (2012).
61. L. Ci, L. Song, Ch. Jin, *et al.*, Atomic layers of hybridized boron nitride and graphene domains, *Nature Materials*, **9**, 430–435 (2010).
62. E.J.H. Lee, K. Balasubramanian, R.T. Weitz, *et al.*, Contact and edge effects in graphene devices, *Nature Nanotech*, **3**, 486–490 (2008).
63. E.J. Kirkland, On the optimum probe in aberration corrected ADF-STEM, *Ultramicroscopy*, **111**, 1523–1530 (2011).
64. W. Zhou, M.P. Oxley, A.R. Lupini *et al.*, Single Atom Microscopy, Microscopy and Microanalysis, (to be published). The graphene sample was obtained from Graphene Supermarket (Available at https://graphene-supermarket.com).
65. M.M.J. Treacy, Z dependence of electron scattering by single atoms into annular dark-field detectors, *Microsc. Microanal.*, **17**, 847–858 (2011).
66. M. Born and E. Wolf, *Principles of Optics*, 7th edn, Cambridge University Press, Cambridge (1997).
67. V. Beck and A.V. Crewe, High resolution imaging properties of the STEM, *Ultramicroscopy*, **1**, 137–144 (1975).
68. R.F. Egerton, *EELS in the Electron Microscope*, 3rd edn, Springer, New York and Heidelberg (2011).
69. R.F. Egerton, Limits to the spatial, energy and momentum resolution of electron energy-loss spectroscopy, *Ultramicroscopy*, **107**, 575–586 (2006).

70. K. Sader, B. Schaffer, G. Vaughan, *et al.*, Smart acquisition EELS, *Ultramicroscopy*, **110**, 998–1003 (2010).
71. F.J. Garcia de Abajo and A. Howie, Electron spectroscopy from outside—aloof beam or near field? *Inst. Phys. Conf. Ser.*, **161**, 327–330 (1999).
72. W. Zhou, M.D. Kapetanakis, M.P. Prange, *et al.*, Direct determination of distinct atomic configurations and bonding of individual impurities in graphene, submitted to *Physical Review Letters* (2012).
73. W. Zhou, J. Lee, J. Nanda, *et al.*, Atomically localized plasmon enhancement in monolayer graphene, *Nature Nanotechnology*, **7**, 161–165 (2012); see also the supporting materials (Available at http://www.nature.com/nnano/journal/vaop/ncurrent/full/nnano.2011.252.html#/supplementary-information: accessed August 5, 2012).
74. W. Zhou, S.J. Pennycook and J.C. Idrobo, Localization of inelastic electron scattering in the low-loss energy regime, *Ultramicroscopy*, **119**, 51–56 (2012). (DOI:10.1016/j.ultramic.2011.11.013.
75. M. Isaacson and D. Johnson, The Microanalysis of light elements using transmitted energy loss electrons, *Ultramicroscopy*, **1**, 33–52 (1975).

7

Low Voltage Scanning Transmission Electron Microscopy of Oxide Interfaces

Robert Klie
Nanoscale Physics Group, Department of Physics, University of Illinois at Chicago, USA

7.1 INTRODUCTION

For over 40 years, electron microscopy, in particular scanning transmission electron microscopy has been used to visualize condensed matter structures with atomic resolution [1]. Beginning in 1970, Crewe *et al.* showed for the first time that using a highly convergent 30 kV electron beam (convergence angle ~20 mrad), single uranium and thorium atoms on 20 nm thick amorphous carbon films can be directly imaged using an annular dark field detector [2]. At the time, the authors reported the smallest possible probe-size of 5 Å, which is sufficient to directly image individual heavy atoms on carbon support, but well in excess of the interatomic spacings in many crystals structures. Atomic-resolution lattice imaging of crystalline materials using annular dark field imaging was not possible using such an electron probe. Following Crewe's initial discovery, over the next few decades the trend in scanning transmission electron microscopy (STEM) moved towards increasing the electron

Low Voltage Electron Microscopy: Principles and Applications, First Edition.
Edited by David C. Bell and Natasha Erdman.

energy from 30 keV up to the 1 MeV regime to decrease the electron wavelength, and therefore also increase the spatial resolution up to the ambitious goal of 0.5 Å in the STEM images [3]. It is interesting to point out here that even at 30 kV, the wavelength of the electron beam is only about 7 pm, that is about 70 times smaller than the reported probe-size by Crewe *et al.* [2] However, aberrations of the probe-forming electron optics, as well as various forms of instabilities seemed to be the limiting factors in achieving significantly smaller electron probe-sizes.

The first commercially available dedicated STEM instruments were the Vacuum Generator (VG) HB5 series first sold in 1974 to Queen Elizabeth College, London, which incorporated a high-brightness cold-field emission electron sources with acceleration voltage of 80 kV (with an overvoltage capacity at 100 kV), and focused on improved mechanical and electronic stability to further increase the imaging power. Eventually, the VG HB5 instrument was operated at 100 kV with an overvoltage capacity at 120 kV. Driven by Crewe's incredible results, other commercial STEMs were developed around that time including the AEI STEM-1 (delivered in 1975), the Siemens Elmskop ST100F (delivered first in 1975 with a total of six instruments built), and a Hitachi STEM (developed in the early 1970s). For a more comprehensive history of dedicated STEM instrumentation, see the recent book by P.W. Hawkes [3].

In the early 1990s, atomic-resolution imaging, as well as atomic-column resolved electron energy-loss spectroscopy (EELS) of bulk crystals, complex interfaces and catalysts first became possible using the VG HB501 and 603, respectively [4–10]. By the late 1990s, the stability of conventional transmission electron microscopy instrumentation had caught up to allow STEM imaging, or better high-angle annular dark field imaging (HAADF), in a conventional TEM instrument operated at 200 keV electron energy with atomic resolution [11–14]. Yet, the lowest possible probe-size that could be achieved in the JEOL2010F at the University of Illinois at Chicago was 1.36 Å for a 200 kV beam, [15, 16] still more than a factor of 50 larger than the electron wavelength.

All this changed in 1997, when the first aberration-corrected STEM instruments demonstrated the possibility of controlling the aberrations of the probe-forming objective lens system using a quadrupole-octupole configuration, and thereby significantly increase the attainable spatial resolution [17–22]. Again, the initial trend was to use higher electron energies to further increase the spatial resolution towards the sub-Å regime. For example, using the aberration-corrected VG603, Nellist *et al.* demonstrated that apparent lattice information can be imaged at 61 pm in a HAADF images of Si [112], the smallest probe-size reported at that time [22]. But even this remarkable probe-size is still more than a factor of 45 larger than the electron wavelength at 300 kV. Again, mechanical and electronic instabilities of the first generation of aberration-corrected instruments, which were initially designed for significantly larger probe-sizes before being retrofitted with aberration correctors, were limiting further progress.

Finally, nearly 10 years after the first custom-made aberration-correctors had been successfully retrofitted on existing instrument columns, of-the-shelf scanning transmission electron microscopes became available, specifically designed for aberration-corrected electron optics and sub-Å spatial resolution [23]. Such instruments were initially designed to be operated at 200 kV or 300 kV, but the flexible designs of the electron optics and the aberration correctors now allow for an instrument operation in a wide range of electron energies, especially in the lower end of the energy spectrum (i.e., 60 or 80 keV) [24]. At the highest electron energy, these newly designed aberration-corrected instruments now provide a spatial resolution has high at 0.5 Å for HAADF imaging and EELS spectroscopy [23, 25].

Moreover, this incredible increase in the instruments' spatial resolution now also allows for much higher beam currents to be used at moderately lower resolution, which provides a significant increase in the signal-to-noise ratio (SNR) of EELS and energy dispersive X-ray spectroscopy (EDS), in particular for the various forms of spectrum imaging. However, it was soon noted that this incredible increase in probe current might also lead to enhanced radiation damage effects in many materials that were considered stable under the electron beam in conventional TEM/STEM instruments.

Guided by the discovery of single layer graphene and boron nitride (BN), the trend finally progressed towards lower primary voltage [26, 27]. With the damage threshold for graphene at about 86 kV (78 kV for BN), the time for low kV imaging and spectroscopy with atomic resolution seemed to have finally come. Many, still ongoing experiments utilize 80 kV, or even 60 kV electron beams to image individual dopants, or vacancies in single layer graphene and BN [27, 28]. The lowest possible probe-size at 60 kV was reported to be 1.2 Å, therefore still small enough to distinguish individual atoms in single layer BN with a B-N distance of 1.45 Å [27]. This probe-size at 60 kV is still smaller than the smallest possible probe-size in a conventional (i.e., uncorrected) TEM/STEM at 200 kV, which are still widely used for atomic-scale characterization of materials. Therefore, with the development of aberration-corrected TEM/STEMs, low kV analysis should also be possible for a wide range of materials systems, not just single-layer structures such as graphene or BN. Nevertheless, graphene and single layer BN exhibit one enormous benefit for low kV analysis in that they are only one monolayer thick. This points to one of the fundamental problems of low kV imaging and spectroscopy of many materials system: sample thickness. Even more so than at higher acceleration voltages, the sample thickness and homogeneity are crucial for any analysis at 60 or 80 kV. Relatively thin surface layers of amorphous carbon can significantly decrease the spatial resolution and SNR of the data, and render any low kV analysis meaningless.

In this chapter, a series of recent experiments performed on well-defined oxide hetero-interfaces using low-kV electron beams using the probe-aberration corrected JEOL ARM200 will be reviewed. More specifically, the ability to simultaneously acquire atomic-resolution *Z*-contrast (or high-angle annular dark field, HAADF) images and annular bright-field (ABF) images will be shown using a $SrTiO_3/BiFeO_3$ interface. Next, the benefits of using 80 and 120 kV electron beams to quantify the ordering of Si atoms at the Si_3N_4/SiO_2 interface will be demonstrated, followed by a study to characterize the effects of oxygen vacancies and Ti diffusion at a $SrTiO_3$/GaAs interface. All the materials systems discussed here are extremely beam sensitive at higher electron energies, and the atomic-resolution analysis reported here would not have been possible without the availability of low kV electron beams in the latest generation of probe corrected TEM/STEMs.

7.2 METHODS AND INSTRUMENTATION

The low-kV characterization of oxide interfaces has been conducted using simultaneous atomic-resolution *Z*-contrast (or HAADF) imaging and annular bright field (ABF), or electron energy-loss spectroscopy (EELS). *Z*-contrast images [9, 29–31] are formed by scanning a highly focused electron beam across the surface of the sample material and collecting the high-angle scattering on an annular dark field detector. Detecting the scattered intensity at these high-angles and integrating over a large angular range effectively averages coherence effects between neighboring atomic columns in the specimen, while thermal vibrations reduce the coherence between atoms in the same column to residual correlations between nearest neighbors [32, 33]. Therefore, in the approximation of large collection angles, each atom can now be considered an independent scatterer with scattering cross-sections approaching $\sim Z^n$, where Z is the atomic number and n ranges between

1.6–2, depending on the collection conditions. At 60 kV and medium angle annular dark field (MAADF) imaging conditions, Krivanek *et al.* have demonstrated that the average atomic column intensity for light atoms (up to $Z = 12$) scales with $Z^{1.64}$ [27]. This cross-section effectively forms an object function that is strongly peaked at the atom sites. Therefore, for very thin specimens where there is no dynamical diffraction, the detected intensity consists of a convolution of this object function with the probe *intensity* profile [34, 35]. For a crystalline material in a zone-axis orientation, where the atomic spacing is greater than the probe size, the atomic columns are illuminated sequentially as the probe is scanned over the specimen and an atomic resolution compositional map is generated. In addition, the phase problem usually associated with the interpretation of conventional high-resolution TEM images is eliminated in Z-contrast imaging.

Z-contrast imaging is therefore highly sensitive to heavy elements and it has been generally accepted that imaging of elements with an atomic number $Z<10$ is nearly impossible. However, annular bright field (ABF) imaging in STEM which utilizes electron scattered within an angular range less than the probe convergence semi-angle can be acquired simultaneously with Z-contrast images in aberration-corrected instruments. It has been demonstrated that this imaging techniques still provides incoherent images without the phase-problem of conventional TEM, [36] but it is also highly sensitive to light elements, including oxygen, nitrogen or even hydrogen [37–39]. Using ABF imaging, the atomic columns of light elements can now be clearly identified, and it might even be possible to directly quantify the occupancy of the individual atomic columns (i.e., assess the vacancy concentrations). While ABF imaging can be used simultaneously with HAADF imaging to provide a complete picture two-dimensional arrangement of all cations and anions in the lattice, ABF cannot be employed at the same time as EELS. Only simultaneous Z-contrast imaging and EELS is possible, since the annular detector used for Z-contrast imaging does not interfere with the low-angle inelastic scattering used for EELS. This means that the Z-contrast image can be used to position the electron probe over a particular structural feature for the acquisition of a spectrum [5–7, 40, 41].

The physical principle behind EELS relates to the interaction of the fast electrons with the sample to cause either collective excitations of electrons in the conduction band, or discrete transitions between atomic energy-levels, for example. $1s \rightarrow 2p$ transitions [42]. Furthermore, the transitions to unoccupied states above the Fermi level allows the degree of hybridization between atomic orbitals to be determined, that is, information on local electronic structure (bonding) changes can be ascertained [42].

A key aspect of the research presented in this chapter is the necessity to be able to correlate the spectrum with a particular structural feature, which requires that the spectra have the same spatial resolution as the Z-contrast image. For crystalline materials in zone-axis orientations, provided a large collection angle (20–50 mrad) is maintained, coherent effects will be averaged and the description of the spectrum in terms of a convolution of the probe with an object function is still valid.

The imaging and spectroscopy analysis reported in this Chapter were performed using the probe-aberration corrected JEOL ARM200F and JEOL ARM200CF at 80 and 120 kV, respectively. While the ARM200F is equipped with a conventional Schottky field emission electron source, the ARM200CF utilizes a cold-field emission source for increased energy resolution during EELS spectroscopy and spectrum imaging. At 120 kV, a probe convergence semi-angle of 29 mrad was used for both imaging and spectroscopy, while an annular dark field (ADF) detector inner angle of 90 mrad and outer angle of 170 mrad was used. Annular bright field (ABF) images were acquired using a detector angle between 14 and 29 mrad at 120 kV. For EELS spectroscopy, a collection semi-angle of 64 mrad was used. At 80 kV, a slightly larger probe convergence semi-angle was used: 30 mrad, with an EELS collection semi-angle of 35 mrad. The detector angles for both HAADF and ABF imaging remained unchanged at 80 kV compared to the 120 kV setting.

7.3 LOW VOLTAGE IMAGING AND SPECTROSCOPY

7.3.1 $SrTiO_3$/$BiFeO_3$ Interface

The first example of low kV analysis on oxide interfaces shown here focuses on 80 kV imaging of a $SrTiO_3$/$BiFeO_3$ interface. Figure 7.1 shows an atomic-resolution Z-contrast and an ABF image of this interface with the $SrTiO_3$ substrate on the left and the $BiFeO_3$ on the right. The atomic structure of the $SrTiO_3$ [001] can be clearly seen with the cubic unit-cell, where the SrO atomic columns are brightest and the TiO_2 are located in the center of the square formed by the SrO columns. It is interesting to note here, that the majority of the $BiFeO_3$ is not oriented along a zone-axis and the lattice is therefore not visible. However, there appears to be a bright band at the $SrTiO_3$/$BiFeO_3$ interface, consisting of 1–2 monolayers of $BiFeO_3$.

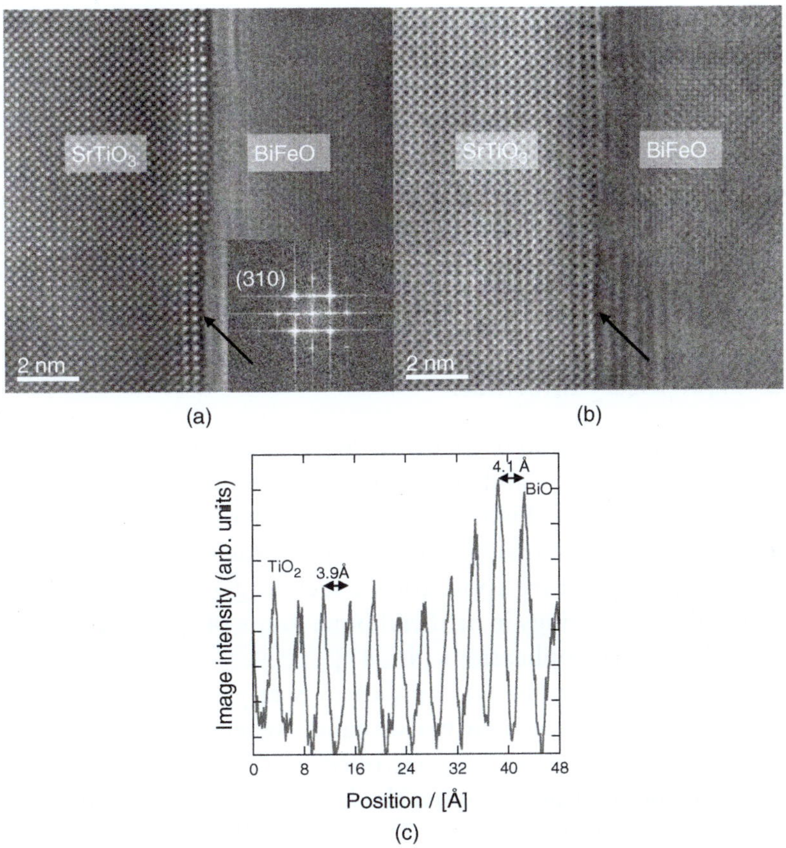

Figure 7.1 Atomic-resolution imaging at 80 kV of a $SrTiO_3/BiFeO_3$ interface. (a) *Z*-contrast and (b) ABF image. (c) Image intensity line profile of the line indicated in (a) showing the TiO_2 and BiO columns. The spacing between the columns is indicated in the image.

The insert in Figure 7.1(a) shows the diffractogram taken from the $SrTiO_3$ support, which clearly shows the (310) spot for $SrTiO_3$, thereby indicating that an electron probe-size as small as 1.2 Å can be achieved at 80 kV. The image further shows a 0.5 monolayer step in the ultra-thin $BiFeO_3$ film (indicated by the arrow). Similarly, Figure 7.1(b) shows the ABF image taken simultaneously with the *Z*-contrast image, which also shows the heavier $BiFeO_3$ layer on the $SrTiO_3$ surface and the steps in the $BiFeO_3$ film. The image intensity line-scan across the hetero-interface is shown in Figure 7.1(c), which shows the TiO_2 columns with an inter-atomic spacing of 3.9 Å (as expected for $SrTiO_3$), while the BiO atomic columns in the two-monolayer thick $BiFeO_3$ film appear to be slightly further apart (i.e., 4.1 Å). This indicates the presence of the tetragonal $BiFeO_3$ phase with $a_t = 3.93$ Å and $c_t = 4.0$ Å.

This example is meant to show that simultaneous atomic-resolution Z-contrast and ABF imaging is possible at a primary voltage as low as 80 kV using conventional TEM sample materials other than graphene or BN. Therefore, beam sensitive materials can now be analyzed with atomic resolution and sub-eV energy resolution (in case of a cold-field emission gun). More detailed studies of a different oxide interfaces using EELS spectrum imaging will be presented next.

7.3.2 Si_3N_4/SiO_2 Interfaces

Silicon nitride ceramics, in particular β-Si_3N_4, are of wide spread interest due to their mechanical and physical properties at room and elevated temperatures, including high resistance to oxidation, low coefficient of frictions and thermal expansion [43, 44]. As such, Si_3N_4 is now used in many high temperature applications, such as gas turbines and engine parts, but also as bearings in artificial bone joints [45] or as thin films in high-speed memory applications [46, 47].

Many of these technological applications rely on the Si_3N_4 crystal/amorphous interfaces for their performance, where the overall mechanical and thermal properties are controlled by the atomic and electronic structures at the interface between the Si_3N_4 grains and an amorphous intergranular film (IGFs) [48, 49]. The very small thickness (1–2 nm) of the IGFs, however, poses difficulties for detailed identification of surface relaxations and reconstructions, interatomic mixing, as well as possible partial ordering at the interface. Some evidence of the latter has emerged, but so far, all have involved heavy cations, which can be easily imaged using high-resolution imaging [50–54]. Key knowledge, however, has been missing concerning the position of light atoms, without which it is not possible to obtain complete understanding of the interfaces. To date, most of the work on IGF films composed of light atoms and their interfaces in Si_3N_4 has focused on atomistic simulations and *ab initio* modeling.

The presence of atomic short-range ordering at the interface between crystalline β-Si_3N_4 and amorphous SiO_2 can be investigated using aberration-corrected scanning transmission electron microscopy (STEM) at low primary voltages, ranging from 80 kV to 120 kV [38]. The images show that first atomic layers of the amorphous SiO_2 film reconstruct in the vicinity of the interface to adopt the crystalline form of Si_3N_4.

These results confirm that the compositional and spatial identification of crystal/amorphous interfaces on the atomic scale has become possible only with the improved spatial resolution of aberration corrected imaging at low voltage.

Annular Bright-field (ABF), and *Z*-contrast imaging, as well as EELS at 120 and 80 kV are used to characterize the interface between Si_3N_4 (1010) and the amorphous SiO_2 IGF at the atomic scale. The grains selected for imaging and spectroscopic measurements were those displaying definite, sharp edges and no significant contrast variation in the vicinity of the Si_3N_4 surface, terminated with open-rings, when viewed edge-on with respect to the electron beam. This ensured that the observed features at the interface with the intergranular film were not due to the surface steps often formed in Si_3N_4 grains.

Figure 7.2 shows atomic-resolution *Z*-contrast and ABF images of bulk Si_3N_4 [0001] acquired at 120 kV primary voltage. A ball-and-stick model of the Si_3N_4 structure is also shown in Figure 7.2(b). Figure 7.2(d) shows the diffractogram of Si_3N_4 bulk, with peaks visible out to the $(8\bar{6}0)$ spot, indicating that at 120kV a probe-size as small as 0.91 Å can be achieved. Using this probe-size, it is now possible to resolve nitrogen

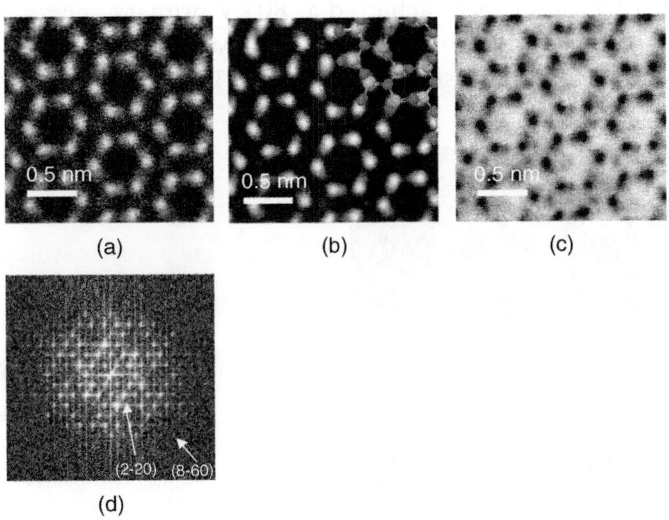

Figure 7.2 Atomic-resolution images of Si_3N_4 (0001) taken at 120 kV: (a) HAADF image; (b) filtered HAADF image and structure of Si_3N_4 overlaid with Si in brown and N in blue; (d) filtered ABF image shown both Si and N atomic columns; (d) diffractogram of Si_3N_4 bulk image showing the (8–60) (See plate section for coloured version).

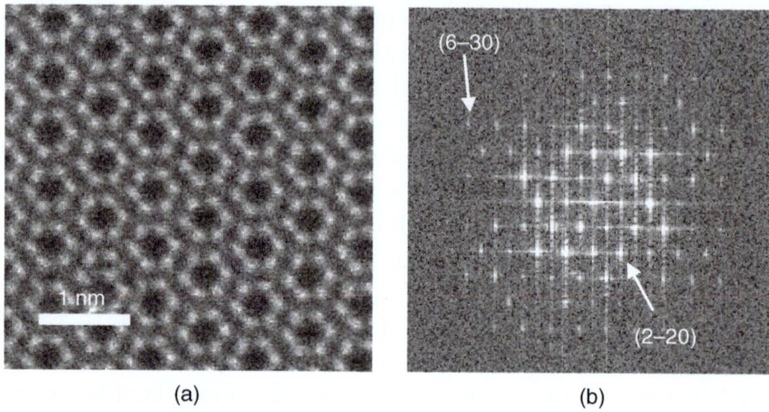

Figure 7.3 (a) Atomic-resolution *Z*-contrast image of Si_3N_4 (0001) at 80 kV; (b) diffractogram of bulk Si_3N_4 showing the (630) peak.

atomic columns even in the Z-contrast images (Figure 7.2a and b), as well as in the ABF image (Figure 7.2c).

Figure 7.3 shows a *Z*-contrast image from a similar area in Si_3N_4 acquired with an 80 kV electron beam. As shown in Figure 7.3(b), the $(6\bar{3}0)$ spots are clearly visible in the diffractogram, indicating that a probe-size of 1.16 Å can be achieved at 80 kV primary energy.

Figures 7.4(a) and 7.4(b) show a *Z*-contrast image and ABF image of the interface acquired simultaneously at 120 kV. The characteristic

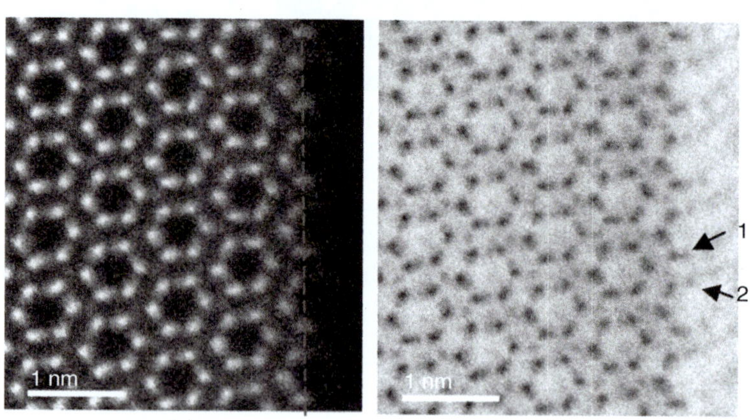

Figure 7.4 Atomic-resolution images of Si_3N_4 [10–10]/SiO_2 interface taken at 120 kV (a) Z-contrast image with the dotted line indicating the Si_3N_4 surface; (b) ABF image marking two different configurations on the Si_3N_4 surface.

hexagonal rings of the ceramic in the [0001] orientation formed by the Si and N atoms are clearly visible in the images. The dotted line marks the termination of the Si_3N_4 surface with the terminating open rings along the (10–10) direction. These open structures have been consistently observed at the Si_3N_4/SiO_2 interfaces.

The ABF image (Figure 7.4b) reveals significant additional information about the interface structure compared to the Z-contrast image. It clearly shows that crystalline order is also present away from the interface, deeper into the amorphous SiO_2 region, as shown by the appearance of dark spots to the right of the dotted line in Figure 7.4(b).

The projected two-dimensional distances from the Si atoms in the terminating rings to the nearest neighbors in the IGF are measured to be approximately (0.097 ± 0.016 nm) and (0.177 ± 0.016 nm). These distances fall in the range of the Si—N and Si—O bond length measurements for the bare and oxygenated Si_3N_4 surfaces, suggesting that the observed atomic arrangements could involve either N and/or O atoms, that is, the observed structures in the amorphous IGF could be partially oxygenated [55]. Although structurally similar, the observed structures in the SiO_2 region are not those of Si_3N_4 closed-ring termination since they are slightly distorted from the ideal closed-ring geometry and they are missing the closing atom on the right of the ring. To identify the composition of the observed structures, atomically-resolved EELS experiments need to be carried out along and across the interface. However, even at 120 kV the Si_3N_4/SiO_2 is too electron beam sensitive to perform any kind of EELS analysis. Therefore, atomic-resolution Spectrum Imaging (SI) in EELS was performed using the JEOL ARM200F at 80 kV with an acquisition time of 0.05 s/pixel.

Figure 7.5 shows the atomic-resolution Z-contrast image taken at 80 kV prior and immediately after the spectrum image acquisition. The area of the spectrum image is indicated by the green box. Judging from these two images, apart from a slight change in focus (due to drift of astigmatism, chroma or defocus), there is no apparent sign of electron beam induced damage to the interface. Figure 7.6(b–e) shows the integrated intensities of the Si L-, N K-, as well as the O K-edges with an integration window of 35 eV. In the case of the Si-L map, the signal integration over a 35 eV window was performed at the edge onset (i.e., 96–131 eV) as well as 200 eV beyond the edge onset. The HAADF signal, acquired simultaneously with the EELS signal is shown in Figure 7.6(a). The schematic drawing of Si_3N_4 crystal structure, with the N atoms in black and the Si atoms in gray, and the terminating open rings are superimposed in the HAADF image. The dark band to the left

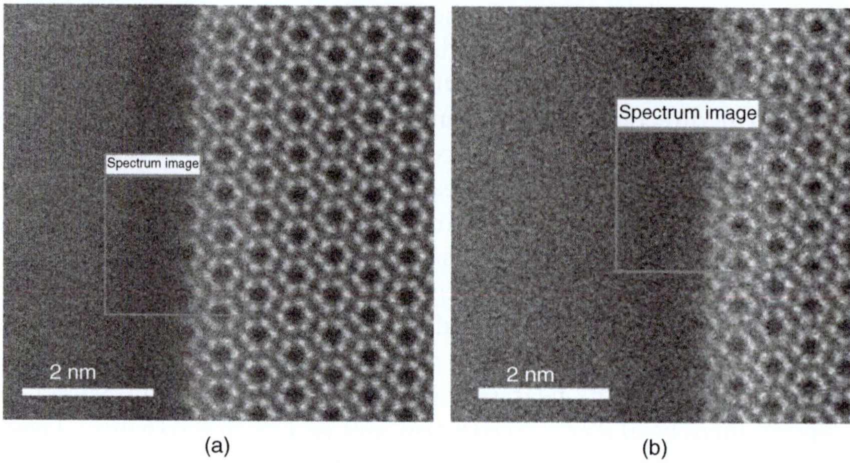

Figure 7.5 Atomic resolution *Z*-contrast image of Si_3N_4/SiO_2 interface with the area of the spectrum image indicated by the green box: (a) before and (b) after the spectrum image acquisition. The total acquisition time was about 5 min.

from the open rings is the SiO_2 layer of approximately 1.4 nm thickness. Figure 7.6 highlights a few interesting features that can only be seen in the spectrum image:

1. *Signal localization.* The Si L-edge integrated signal appears to be delocalized at the edge onset, while the integrated intensity well beyond the edge onset shows a much more localized integrated intensity profile, compared to the N K-edge signal. The calculated delocalization factors of the Si L_{23} and the N-K edges at 80 kV are 0.53 nm and 0.2 nm, respectively, [42] while the diagonal and the nearest interatomic distances of the Si columns are 0.56 nm and 0.31 nm. Therefore, the delocalization of the Si L_{23}-edge is larger than the interatomic distance between the neighboring Si atoms, and the Si signal is expected to be delocalized. Wang *et al.* have previously shown in this case one should place the energy window as far above ionization threshold as the background subtraction procedure justifiable allows. Figure 7.6(c) shows the Si core-loss image obtained with an energy window from 300–335 eV, and the Si atomic columns now become visible. This result is in agreement with the observations of Kimoto *et al.*, who studied the localization of elastic and inelastic scatterings using bulk Si_3N_4 samples, but with different image acquisition conditions [56].

2. *Effect of beam damage*. The individual elemental maps in Figure 7.6 clearly show that there is no nitrogen present in the SiO_2 films, but there appears to be a significant amount of sub-surface oxygen in the Si_3N_4 grains. At the interface, the oxygen signal is mostly concentrated in between the open rings. However, in the lower portion of the image, the signal is found to go much deeper into the Si_3N_4 region, as far as 5 Å, resulting in an apparently much wider SiO_2 layer. Looking at the corresponding regions in the N and Si maps, one can see that oxygen is probably replacing nitrogen, as Figure 7.6(d) shows a decrease in the intensity at the location where the increase in the O signal intensity is observed. Although energetically favorable, [55] the O→N replacement in the vicinity of the bottom ring is, most likely, the effect of electron beam damage due to the prolonged exposure of the sample during the spectrum image acquisition. It is however interesting to note that such beam induced damage is not (yet) visible in the Z-contrast image (Figure 7.5b).

7.3.3 Ultrathin $SrTiO_3$ films on GaAs

In the past years, ultrathin metal-oxide films on polar semiconductor surfaces have received much attention due to the occurrence of novel interfacial properties, including ferroelectricity, superconductivity and the presence of an interfacial two-dimensional electron gas [57–62]. It has been more than two decades since the growth of such systems started, and various techniques have been used to approach a layer-by-layer crystalline oxide film and avoid the formation of the amorphous layer at the interface [63, 64].

The $SrTiO_3$/GaAs interfaces were studied experimentally and theoretically with various techniques in order to develop a fundamental mode of the structure property relationships and it has been reported that the $SrTiO_3$ film prefers to be SrO terminated at the interface, regardless of the growth condition, and a submonolayer of Ti between the oxide and semiconductor can release the Fermi level pinning after the thin film deposition [65–67]. However, due to the oxygen deficiency of the film, it was not possible to perform atomic resolution electron energy-loss spectroscopy (EELS) on the highly beam sensitive $SrTiO_3$/GaAs interface [65].

Figure 7.7(a) shows an atomic-resolution Z-contrast image of a four-monolayer thin $SrTiO_3$ film on GaAs (001), where the epitaxy is apparent

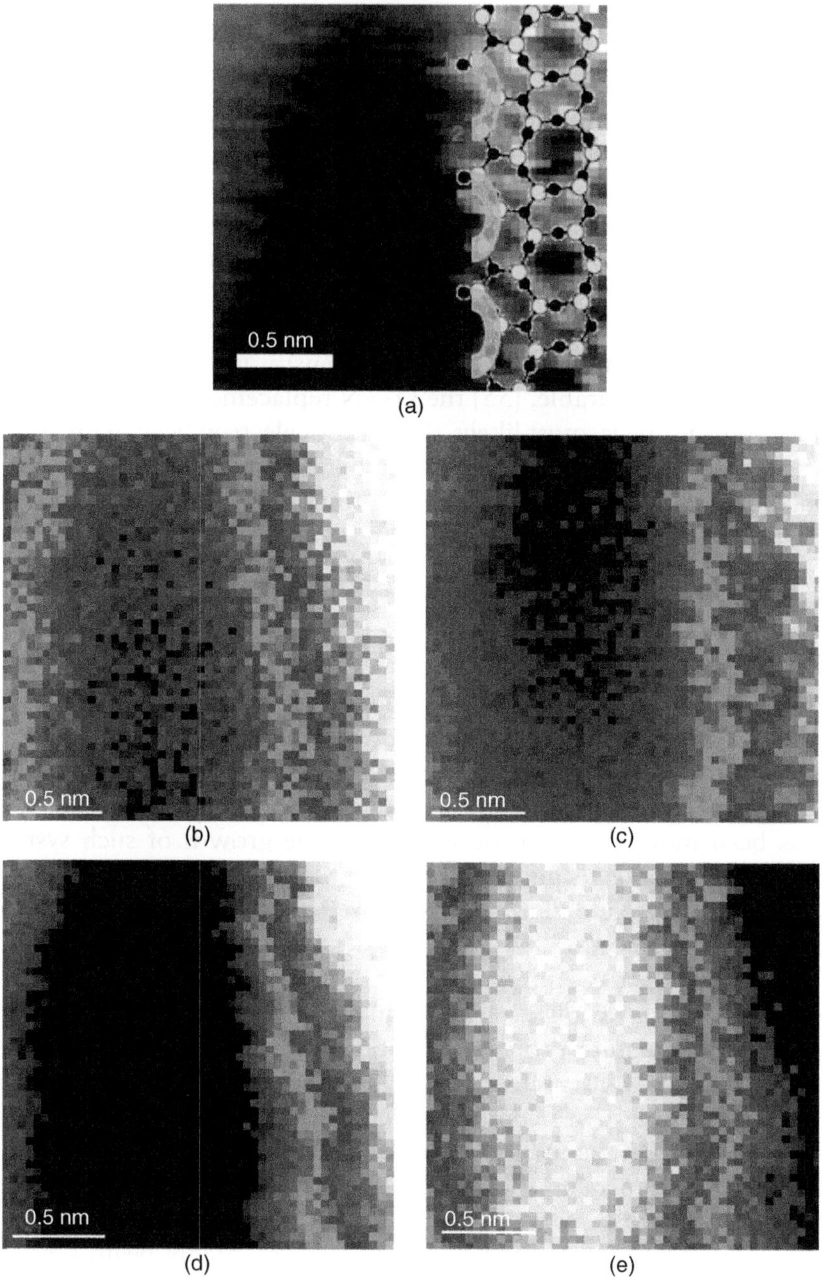

Figure 7.6 Spectrum imaging of Si_3N_4/SiO_2 interface at the atomic scale: (a) HAADF signal acquired simultaneously with the spectrum image shown in (b–e); (b) integrated Si L-edge intensity; (c) integrated extended L-edge fine-structure (200 eV above the edge onset); (d) N K-edge integrated intensity; (e) O K-edge integrated intensity (See plate section for coloured version).

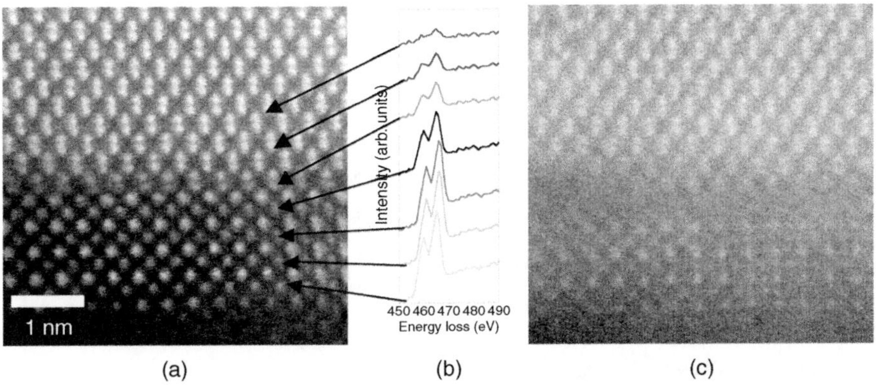

Figure 7.7 Atomic-resolution Z-contrast image of $SrTiO_3$/GaAs. (a) After a single image scan across the field of view. (b) Ti L-edge taken from the position indicated in (a), showing clear signs of Ti diffusion into the top-layers of the GaAs support. (c) Same region as shown in (a) after EELS measurements. The interfacial structure and the $SrTiO_3$ show signs of severe beam damage. The electron energy for this experiment was 100 keV (See plate section for coloured version).

with $SrTiO_3$(001) || GaAs (001) and $SrTiO_3$[110] || GaAs[100]. The interface between the $SrTiO_3$ film and the GaAs-support in Figure 7.7(a) appears sharp with no obvious interfacial diffusion. Moreover, the image suggests that the oxide film starts with a SrO layer at the As terminated GaAs interface. However, the $SrTiO_3$ films exhibit high sensitivity to the electron beam at both 300 kV and 100 kV (Figure 7.7c), which has made it nearly impossible to achieve reliable EELS quantification. Figure 7.7(b) shows the Ti L-edge taken from the position indicated in Figure 7.7(a) taken at 100 kV primary energy with an exposure time of 0.2 s per pixel. Figure 7.7(c), which was acquired immediately after the EELS analysis, shows that even at 100 kV and with very short exposure times, the $SrTiO_3$/GaAs interface has been severely damaged by the electron beam, thereby making any detailed analysis impossible.

By further decreasing the electron energy, more specifically, by using an 80 kV electron beam in the JEOL ARM200CF, the $SrTiO_3$/GaAs is now sufficiently stable to withstand atomic resolution imaging and spectroscopy for extended periods of time. Even several hours of analysis on a particular region of a sample has so far not shown any signs of beam damage, neither in Z-contrast imaging nor in spectroscopy.

The acquired Ti L- and O K-edges are shown in Figure 7.8, respectively. The top spectrum in Figure 7.8(a) is taken from a $SrTiO_3$ bulk specimen thus providing a Ti^{4+} fine structure as a reference. The crystal field splitting of Ti L_3 and L_2 edges can be clearly resolved in

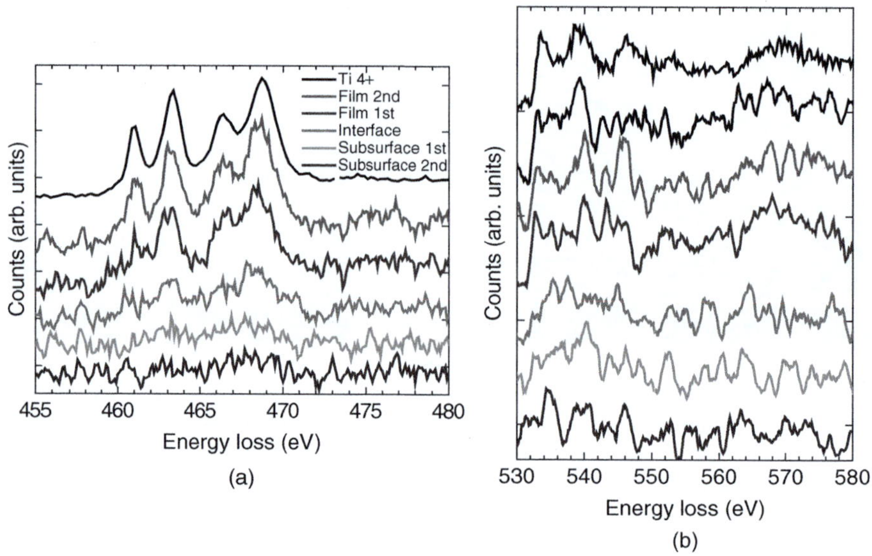

Figure 7.8 (a) Titanium L-edge and (b) oxygen K-edges across the $SrTiO_3$/GaAs interface. The location of the spectra is indicated by the colored rectangle in Figure 7.8 (See plate section for coloured version).

all the spectra taken from the $SrTiO_3$ thin film, which indicates that the Ti valence is close to 4+ throughout the film [68]. To further assess the Ti valence state, and thereby the oxygen stoichiometry of the films, the energy scale has been calibrated using the O K-edge onset, as shown in Figure 7.8(b). It can now be seen that the Ti L-edges for each location in the films is slightly shifted towards lower energy, again indicating a slight decrease in the Ti More specifically, both L_3 and L_2 edges shift to lower energies as the electron probe approaches to the $SrTiO_3$/GaAs interface, indicating a decrease of the Ti valence from 4+ to a mixture state of 3+ and 4+ [68]. Such phenomena could be caused by oxygen vacancies, or because the Ti-O bonding state on the TiO_2 columns in the thin film changed to Ti-As on the surface of the substrate.

In addition, it is interesting to note here that a Ti signal is noticeable at least two layers into the GaAs support, indicating that some of the Ti has diffused into the GaAs support during the film synthesis. This results confirms earlier measurements (shown in Figure 7.7) using a 100 kV beam.

In Figure 7.8(b), the acquired O K edges are fitted to their fifth nearest neighbors and compared with the reference spectrum taken from bulk $SrTiO_3$. The colored spectra correspond to the colored rectangles in

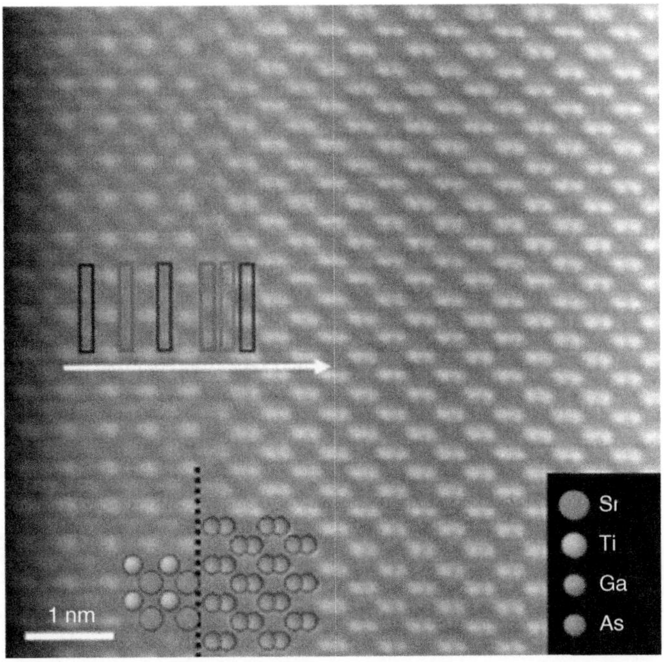

Figure 7.9 Atomic resolution Z-contrast image taken at 80 kV of $SrTiO_3$/GaAs interface. The proposed interfacial structure is shown in the ball-and-stick model, while the colored rectangles indicated the location of the acquired EELS spectra (See plate section for coloured version).

Figure 7.9. It is immediately noticeable that the O K-edge fine-structure in the $SrTiO_3$ thin films is significantly different from the bulk $SrTiO_3$. More specifically, the pre-peak, which strongly scales, with the presence of oxygen vacancies, is suppressed in the spectra taken from the films, [14, 69] indicating the presence of oxygen vacancies. Furthermore, near the $SrTiO_3$/GaAs interface, the fine-structure of the O K-edge exhibits several peaks that were not found in the bulk $SrTiO_3$, or in the films $SrTiO_3$ spectra. These additional peaks could indicate the oxygen bonding with arsenic on the GaAs surface during the initial steps of the thin film synthesis. Finally, the integrated O K-edge intensity disappears completely in the GaAs support, even in the locations where a Ti signal was found. This further indicates that only the Ti diffuses into the support, while the oxygen remains in the $SrTiO_3$ film and on the GaAs surface. Most importantly, the EELS analysis does not exhibit any signs of beam induced damage or oxygen diffusion, as previously observed in the Si_3N_4/SiO_2 interface.

7.4 SUMMARY

In summary, the research presented in this Chapter demonstrates that atomic-resolution analysis, using Z-contrast imaging in combination with ABF imaging or EELS, has become possible with the successful integration of probe-forming aberration correctors into TEM/STEM instruments. Using a primary energy of 80 kV, beam sensitive samples, in particular oxide hetero-interfaces, can now be studied, and their atomic and electronic structures can be explored. As shown for the Si_3N_4/SiO_2 interfaces, the effects of beam induced damage beam are not always immediately apparent in the Z-contrast image even at 80 kV, and only a careful analysis of the associated EELS spectra can reveal these effects. It is, however, not clear if a further decrease in primary voltage beyond 60 kV is beneficial, since the sample thickness of these polycrystalline materials might become the limiting factor.

Future developments of in-situ capabilities for low-kV imaging will be the next great step in this relatively young area of atomic-resolution low-kV imaging and analysis in STEM. Effects of in-situ heating or cooling on the local atomic and electronic structures can then be studied even for beam sensitive samples.

ACKNOWLEDGEMENTS

This work was supported with funding from the National Science Foundation (DMR-0846748, DMR-0605964) and the US Army Research Office (W911NF-10-1-0147). The UIC JEOL JEM-ARM200CF is supported by a MRI-R^2 grant from the National Science Foundation (Grant No. DMR-0959470). The author acknowledges support from the RRC-East in operating the UIC JEOL JEM-ARM200CF.

REFERENCES

1. A.V. Crewe, Scanning Electron Microscopes: Is High Resolution Possible? *Science*, **154**, 729–738 (1966).
2. A.V. Crewe, J. Wall, and J. Langmore, Visibility of Single Atoms, *Science*, **168**, 1338–1340 (1970).
3. P.W. Hawkes, Cold Field Emission and the Scanning Transmission Electron Microscope, in *Advances in Imaging and Electron Physics*, **159**, Academic Press, (2009).
4. S.J. Pennycook, Atomic-scale imaging of materials by Z-contrast scanning-transmission electron microscopy, *Anal. Chem.*, **64**, A263–A272 (1992).

5. N.D. Browning, M.F. Chisholm, and S.J. Pennycook, Atomic-Resolution Chemical-Analysis Using a Scanning-Transmission Electron-Microscope, *Nature*, **366**, 143–146 (1993).
6. P.E. Batson, Simultaneous STEM Imaging and Electron-Energy-Loss Spectroscopy with Atomic-Column Sensitivity, *Nature*, **366**, 727–728 (1993).
7. D.A. Muller, Y. Tzou, R. Raj, and J. Silcox, Mapping sp^2 and sp^3states of carbon at subnanometer spatial-resolution, *Nature*, **366**, 725–727 (1993).
8. P.D. Nellist and S.J. Pennycook, Direct imaging of the atomic configuration of ultradispersed catalysts, *Science*, **274**, 413–415 (1996).
9. S.J. Pennycook and L.A. Boatner, Chemically sensitive structure-imaging with a scanning-transmission electron-microscope, *Nature*, **336**, 565–567 (1988).
10. S.J. Pennycook and D.E. Jesson, High-resolution incoherent imaging of crystals, *Phys. Rev. Lett.*, **64**, 938–941 (1990).
11. E.M. James, Y. Xin, and N.D. Browning, Optimizing the STEM performance of the JEOL 2010F 200kV FEGTEM, in *Electron Microscopy and Analysis 1999*, IOP Publishing Ltd, Bristol (1999).
12. K. Kishida and N.D. Browning, Atomic resolution Z-contrast imaging of grain boundaries in Ag-sheathed BSCCO high-T_c superconducting tapes, in *Electron Microscopy and Analysis 1999*, IOP Publishing Ltd: Bristol (1999).
13. E.M. James and N.D. Browning, Atomic resolution scanning transmission electron microscopy on the 200 kV FEGTEM, *Scanning*, **21**, 91–92 (1999).
14. R.F. Klie and N.D. Browning, Atomic scale characterization of oxygen vacancy segregation at $SrTiO_3$ grain boundaries, *Appl. Phys. Lett.*, 77, 3737–3739 (2000).
15. E.M. James, and N.D. Browning, Practical aspects of atomic resolution imaging and analysis in STEM, *Ultramicroscopy*, **78**, 125–139 (1999).
16. E.M. James, N.D. Browning, A.W. Nicholls, *et al.*, Demonstration of atomic resolution Z-contrast imaging by a JEOL JEM-2010F scanning transmission electron microscope, *J. Electr. Microsc.*, **47**, 561–574 (1998).
17. M. Haider, S. Uhlemann, and J. Zach, Upper limits for the residual aberrations of a high-resolution aberration-corrected STEM, *Ultramicroscopy*, **81**, 163–175 (2000).
18. N. Dellby, O.L. Krivanek, P.D. Nellist, P.E. Batson, and A.R. Lupini, Progress in aberration-corrected scanning transmission electron microscopy, *J. Electr. Microsc.*, **50**, 177–185 (2001).
19. P.E. Batson, N. Dellby, and O.L. Krivanek, Sub-angstrom resolution using aberration corrected electron optics, *Nature*, **419**, 94–94 (2002).
20. L.J. Allen, S.D. Findlay, A.R. Lupini, *et al.*, Atomic-resolution electron energy loss spectroscopy imaging in aberration corrected scanning transmission electron microscopy, *Phys. Rev. Lett.*, **91**, 105503–105506 (2003).
21. O.L. Krivanek, P.D. Nellist, N. Dellby, *et al.* gyi, Towards sub-0.5 angstrom electron beams, *Ultramicroscopy*, **96**, 229–237 (2003).
22. P.D. Nellist, M.F. Chisholm, N. Dellby, *et al.*, Direct sub-angstrom imaging of a crystal lattice, *Science*, **305**, 1741–1741 (2004).
23. T. Isabell, J. Brink, M. Kawasaki, *et al.*, Development of a 200kV Atomic REsolution Analytical Electron Microscope, *Microsc. Today*, **17**, 8–11 (2009).

24. B. Freitag, G. Knippels, S. Kujawa, *et al.*, First performance measurements and application results of a new high brightness Schottky field emitter for HR-S/TEM at 80–300kV acceleration voltage, in *European Microscopy Congress* (2008).
25. C. Kisielowski, B. Freitag, M. Bischoff, *et al.*, Detection of single atoms and buried defects in three dimensions by aberration-corrected electron microscope with 0.5-angstrom information limit, *Microsc. and Microanal.*, **14**, 469–477 (2008).
26. C.O. Girit, J.C. Meyer, R. Erni, *et al.*, Graphene at the edge: stability and dynamics, *Science*, **323**, 1705–1708 (2009).
27. O.L. Krivanek, M.F. Chisholm, V. Nicolosi, *et al.*, Atom-by-atom structural and chemical analysis by annular dark-field electron microscopy, *Nature*, **464**, 571–574 (2010).
28. O.L. Krivanek, N. Dellby, M.F. Murfitt, *et al.*, Gentle STEM: ADF imaging and EELS at low primary energies, *Ultramicroscopy*, **110**, 935–945 (2010).
29. R.F. Loane, P. Xu, and J. Silcox, Incoherent imaging of zone axis crystals with ADF STEM, *Ultramicroscopy*, **40**, 121–138 (1992).
30. S. Hillyard, R.F. Loane, and J. Silcox, Annular Dark-Field Imaging—Resolution And Thickness Effects, *Ultramicroscopy*, **49**, 14–25 (1993).
31. A. Amali, and P. Rez, Theory of lattice resolution in high-angle annular dark-field images, *Microsc. and Microanal.*, **3**, 28–46 (1997).
32. D.E. Jesson and S.J. Pennycook, Incoherent imaging of thin specimens using coherently scattered electrons, in *Proceedings of The Royal Society of London Series A—Mathematical Physical and Engineering Sciences*, **441**, 261–281 (1993).
33. D.E. Jesson and S.J. Pennycook, Incoherent imaging of crystals using thermally scattered electrons, in *Proceedings of The Royal Society of London Series A-Mathematical and Physical Sciences*, **449**, 273–293 (1995).
34. P.D. Nellist and S.J. Pennycook, SubAngstrom resolution imaging using annular dark-field STEM, in *Electron Microscopy And Analysis 1999*, IOP Publishing Ltd, Bristol (1999).
35. P.D. Nellist, and S.J. Pennycook, Incoherent imaging using dynamically scattered coherent electrons, *Ultramicroscopy*, **78**, 111–124 (1999).
36. S.D. Findlay, N. Shibata, H. Sawada, *et al.*, Robust atomic resolution imaging of light elements using scanning transmission electron microscopy, *Appl. Phys. Lett.*, **95**, 191913–191915 (2009).
37. S.D. Findlay, N. Shibata, H. Sawada, *et al.*, Dynamics of annular bright field imaging in scanning transmission electron microscopy, *Ultramicroscopy*, **110**, 903–623 (2010).
38. W. Walkosz, R.F. Klie, S. Ogut, *et al.*, Crystal-induced effects at crystal/amorphous interfaces: The case of Si_3N_4/SiO_2, *Phys. Rev. B*, **82**, 081412 (2010).
39. R. Ishikawa, E. Okunishi, H. Sawada, *et al.*, Direct imaging of hydrogen-atom columns in a crystal by annular bright-field electron microscopy, *Nature Materials*, **10**, 278–281 (2011).
40. M. Varela, S.D. Findlay, A.R. Lupini, *et al.*, Spectroscopic imaging of single atoms within a bulk solid, *Phys. Rev. Lett.*, **92**, 095502 (2004).
41. R.F. Klie and Y. Zhu, Atomic resolution STEM analysis of defects and interfaces in ceramic materials, *Micron*, **36**, 219–231 (2005).

42. R.F. Egerton, *Electron Energy Loss Spectroscopy in the Electron Microscope*. 2nd edn, Plenum Press, New York (1986).
43. I.W. Chen, P.F. Becher, M. Mitomo, *et al.*, Silicon nitride ceramics scientific and technological advances, in *Materials Research Society* (1993).
44. M.J. Hoffmann, Analysis of microstructural development and mechanical properties of Si_3N_4 ceramics, in *Tailoring of Mechanical Properties of Si_3N_4 Ceramics*, M.J. Hoffmann and G. Petzow (eds), Kluwer Academic Publishers, Dordrecht (1994).
45. G. Willmann, Improving bearing surfaces of artificial joints, *Adv. Eng. Mat.*, **3**, 135–141 (2001).
46. M. Saraf, R. Akhvlediani, R. Edrei, *et al.*, Low thermal budget $SiO_2/Si_3N_4/SiO_2$ stacks for advanced SONOS memories, *J. Appl. Phys.*, **102**, 054512 (2007).
47. M. Saraf, R. Edrei, R. Shima-Edelstein, *et al.*, Nitrogen diffusion and accumulation at the Si/SiO_2 interface in $SiO_2/Si_3N_4/SiO_2$ structures for nonvolatile semiconductor memories, *J. Vac. Sci. Technol. B*, **23**, 1558–1561 (2005).
48. D.R. Clarke, On the equilibrium thickness of intergranular glass phases in ceramic materials, *J. Am. Ceram. Soc.*, **70**, 15–22 (1987).
49. R.L. Satet and M.J. Hoffmann, Influence of the rare-earth element on the mechanical properties of RE-Mg-bearing silicon nitride, *J. Am. Ceram. Soc.*, **88**, 2485–2490 (2005).
50. N. Shibata, S.J. Pennycook, T.R. Gosnell, *et al.*, Observation of rare-earth segregation in silicon nitride ceramics at subnanometre dimensions, *Nature*, **428**, 730–733 (2004).
51. A. Ziegler, J.C. Idrobo, M.K. Cinibulk, *et al.*, Interface structure and atomic bonding characteristics in silicon nitride ceramics, *Science*, **306**, 1768–1770 (2004).
52. A. Ziegler, J.C. Idrobo, M.K. Cinibulk, *et al.*, Atomic-resolution observations of semicrystalline intergranular thin films in silicon nitride, *Appl. Phys. Lett.*, **88**, 041919 (2006).
53. G.B. Winkelman, C. Dwyer, T.S. Hudson, *et al.*, Three-dimensional organization of rare-earth atoms at grain boundaries in silicon nitride, *Appl. Phys. Lett.*, **87**, 061911 (2005).
54. G.B. Winkelman, C. Dwyer, C. Marsh, *et al.*, The crystal/glass interface in doped Si_3N_4, *Mat. Sci. Engineer. A*, **422**, 77–84 (2006).
55. W. Walkosz, J.C. Idrobo, R.F. Klie, and S. Ogut, Reconstructions and nonstoichiometry of oxygenated beta-Si_3N_4 (10-10) surfaces, *Phys. Rev. B*, **78**, 165322 (2008).
56. K. Kimoto, K. Ishizuka, and Y. Matsui, Decisive factors for realizing atomic-column resolution using STEM and EELS, *Micron*, **39**, 257–262 (2008).
57. N.A. Pertsev, A.K. Tagantsev, and N. Setter, Phase transitions and strain-induced ferroelectricity in $SrTiO_3$ epitaxial thin films, *Phys. Rev. B*, **61**, R825–R829. (2000).
58. C.H. Ahn, K.M. Rabe, and J.M. Triscone, Ferroelectricity at the nanoscale: Local polarization in oxide thin films and heterostructures, *Science*, **303**, 488–491 (2004).
59. J.H. Haeni, P. Irvin, W. Chang, *et al.*, Room-temperature ferroelectricity in strained $SrTiO_3$, *Nature*, **430**, 758–761 (2004).

60. J.C. Woicik, H. Li, P. Zschack, *et al.*, Anomalous lattice expansion of coherently strained $SrTiO_3$ thin films grown on Si(001) by kinetically controlled sequential deposition, *Phys. Rev. B*, 73, 024112 (2006).
61. M.P. Warusawithana, C. Cen, C.R. Sleasman, *et al.*, A ferroelectric oxide made directly on silicon, *Science*, **324**, 367–370 (2009).
62. A. Ohtomo, and H.Y. Hwang, A high-mobility electron gas at the $LaAlO_3$/$SrTiO_3$ heterointerface, *Nature*, **427**, 423–426 (2004).
63. E.J. Tarsa, M. Degraef, D.R. Clarke, *et al.*, Growth and characterization of (111) and (001) oriented MgO films on (001) GaAs, *J. Appl. Phys.*, **73**, 3276–3283 (1993).
64. M.G. Blamire, J.L. MacManus-Driscoll, N.D. Mathur, and Z.H. Barber, The materials science of functional oxide thin films, *Adv. Mat.*, **21**, 3827–3839 (2009).
65. R.F. Klie, Y. Zhu, E.I. Altman, and Y. Liang, Atomic structure of epitaxial $SrTiO_3$–GaAs(001) heterojunctions, *Appl. Phys. Lett.*, **87**, 143106 (2005).
66. Y. Liang, J. Curless, and D. McCready, Band alignment at epitaxial $SrTiO_3$-GaAs(001) heterojunction, *Appl. Phys. Lett.*, **86**, 082905 (2005).
67. Y. Liang, J. Kulik, T.C. Eschrich, *et al.*, Hetero-epitaxy of perovskite oxides on GaAs(001) by molecular beam epitaxy, *Appl. Phys. Lett.*, **85**, 1217–1219 (2004).
68. D.A. Muller, N. Nakagawa, A. Ohtomo, *et al.*, Atomic-scale imaging of nano-engineered oxygen vacancy profiles in $SrTiO_3$, *Nature*, **430**, 657–661 (2004).
69. N.D. Browning, M.F. Chisholm, S.J. Pennycook, *et al.*, Correlation between hole depletion and atomic-structure at high-angle grain-boundaries in $YBa_2Cu_3O_{7-\delta}$, *Physica C*, **212**, 185–190 (1993).

8

What's Next? The Future Directions in Low Voltage Electron Microscopy

David C. Bell[1] and Natasha Erdman[2]
[1] *School of Engineering and Applied Sciences, Harvard University, USA*
[2] *JEOL USA Inc., USA*

8.1 INTRODUCTION

With the advances in electron gun design, aberration corrected optics and improvements in the sensitivity of recording images and obtaining analytical information, the future for low voltage electron microscopy is looking very promising. Rather than having custom designed SEMs and TEMs, the experience with lower voltage instruments have motivated instrument manufacturers to include these advances with the design of traditional systems. For example, the advent and incorporation of special aberration correctors required improvements in high voltage power supplies and electron optics, these improvements have typically been carried over to non-corrected systems.

In the case of SEMs, the improved operation at primary beam voltages below 1 kV, for example, now means that most FEG-SEM's can easily operate at 0.1–0.5 kV with good performance. Some may say that the catch up is now being done with the detectors to improve upon the efficiency of SE and BSE detection and analytical capabilities. TE and STEM have of course benefited greatly from the incorporation

Low Voltage Electron Microscopy: Principles and Applications, First Edition.
Edited by David C. Bell and Natasha Erdman.

of spherical aberration correctors; now the next step is taking place, the detector design and performance improvements. For TEM the big push is to replace CCD CMOS technology for the imaging array with direct electron detectors, and to incorporate either electron monochromators or chromatic aberration correctors. Limitations of TEMs will then approach the theoretical electron optical resolution as primarily determined by the wavelength. The mantra for low voltage electron microscopy continues to be, "make every electron count" and the continued improvements will ensure this is continued.

The limiting factor for the ultimate resolution of electron microscope is the electron wavelength; moreover, the actual working resolution of electron microscopes is directly limited by the lens aberrations present, as was highlighted throughout this volume. For a 200 kV electron the wavelength is 2.5 pm, but for a 40 kV electron the wavelength becomes nearly three times as large at 6 pm. For an SEM operating at 1 kV the wavelength becomes 39 pm, still a small value but a significantly larger wavelength as compared to, for example, 40 kV.

This chapter will touch upon several design approached and directions for low voltage SEM and TEM that have become prominent in the recent years and will likely continue to have an impact on how the users and instrument manufacturers view imaging and analysis at low voltages. Further advances in other optics, such as ion optics, bring the possibility of different gaseous ion sources being used for imaging and materials fabrication, providing the user with imaging results that are complimentary to the low voltage SEM imaging, and in some cases surpass the more traditional approaches.

8.2 UNIQUE LOW VOLTAGE SEM AND TEM INSTRUMENTS

8.2.1 Miniature SEM Columns

In Chapters 2 and 6 of this volume we highlighted the recent developments in the traditional column and filed emission source design for improved SEM and TEM performance at low voltages. However, no discussion would be complete without mentioning several unconventional electron column designs that have been gaining some prominence in the recent years.

Miniaturized low voltage SEM columns have been first proposed by Chang *et al.* (1990), and have been transformed into a commercial instrument by Novelx (currently part of Agilent Technologies). The entire column measures only 1 cm in length and is completely electrostatic. The column is utilized exclusively for low voltage (up to 2 kV), and has resolution of sub 10 nm at 1 kV. The idea behind miniaturization of the SEM column is the ability to reduce lens aberrations and thus improve low voltage performance by scaling down the size of the lens components (Chang *et al.*, 1990; Murray, 2011). The individual miniature lens components are micromachined using silicon microfabrication techniques. Each column consists of a Schottky FE source, condenser and objective lenses (with no crossovers), high speed blanker and octopole deflector. The signal detection is achieved by utilizing an annular microchannel plate (MCP) detector. Full details of the fabrication and alignment procedure can be found elsewhere (Muray *et al.*, 2006; Spallas *et al.*, 2006). The unique column design achieves substantially reduced value for chromatic aberration at low voltages ($C_c = 2.9$ mm), largely due to the reduced focal length of the miniature lens and short working distance.

There are several advantages to utilization of miniature electron columns. The silicon based components of the system along with the small size of the column (and the overall instrument) lead to low overall operating costs, since the instrument requires only nominal facilities (vacuum, power). The miniature size of the column also eliminates the need for extensive electromagnetic shielding and alleviates concerns of potential mechanical instabilities. Moreover, the individual columns can also be assembled into a microarray for either high-throughput imaging or lithography applications (Murray *et al.*, 2006).

8.2.2 Dedicated Low Voltage TEM

Another unconventional system that should be mentioned is a dedicated low voltage TEM (LVEM5, Delong Instruments, Czech Republic) that was developed several decades ago (Delong, 1992; Delong *et al.*, 1994; Delong *et al.*, 2000) and has been shown to provide new and interesting contrast mechanisms for imaging of different biological and polymer systems (Coufalova and Delong, 2000; Sharp *et al.*, 2000; Lednický *et al.*, 2000). LVEM5 instrument operates at 5 kV accelerating voltage and is composed of two-microscope system: electron and light microscopes (see Figure 8.1). The electron microscope portion features a Schottky FEG source, condenser and objective lenses with permanent magnets, electrostatic intermediate projection lenses, as well as single crystal YAG (Yttrium Aluminum Garnet) fluorescent screen. The small

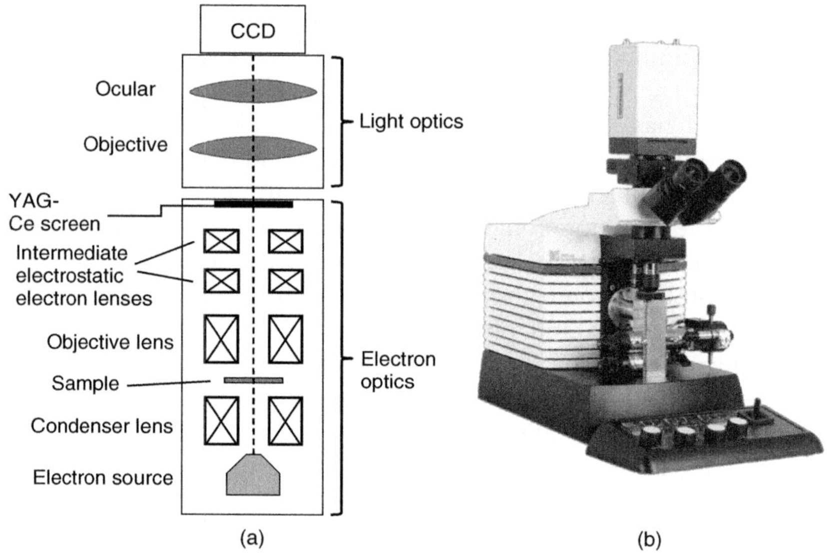

Figure 8.1 (a) Schematic of the low voltage TEM (LVEM5, Delong Instruments); (b) LVEM5 (courtesy of Delong Instruments).

electron microscope generates an image that is converted into a light microscope image by utilizing the YAG crystal. The objective lens of the light microscope has a high numerical aperture (NA = 0.9) and magnifies the image up to 400 times, allowing efficient observation of the specimen. Although this microscope is the smallest TEM available, it features all the imaging modes accessible on the more traditional transmission microscopes, including TEM, electron diffraction, SEM (backscatter) and STEM modes, capable of observing the specimen with few nanometers spatial resolution. The advantage of the LVEM5 is the ability to examine beam sensitive specimens that are comprised of low atomic number elements, such as biological, organic or polymer materials, without a need for additional staining, shadowing or other preparation techniques. Decreasing the accelerating voltage from 120 kV (traditionally used for organic samples observation in TEM) to 5 kV improves the observed contrast almost 10-fold. More recently, utilization of 5 kV LVEM5 revealed existence of pancreatic acinar cell secretory granules with non-homogeneous matrix and sub-compartments having circular profiles that were not previously detected by conventional 80 kV TEM (Bendayan *et al.*, 2011). The microscope was also shown to provide excellent imaging contrast with minimal damage as compared to the traditional TEM when imaging polymer structures, such as pentacene and triblock copolymer films and electrospun polymer fibers (Drummy *et al.*, 2004). Even though sample preparation required for LVEM5 is minimal as compared to the traditional methods, the contrast mechanisms as well as image interpretation (especially in the transmission modes) depend significantly on the specimen thickness and care should be taken when trying to attribute observed features to either elemental or thickness-related differences in the sample.

8.2.3 The Helium Ion Microscope as an Alternative to Low Voltage SEM Imaging

Current research in the areas of nanomaterials and nanobiology requires the corresponding ultra-high resolution imaging techniques that allow examination of the specimens with minimal damage. The technique of choice for such investigations has historically been the field emission SEM. As we have discussed in earlier chapters, the probe size is on the order of 1 nm in FE-SEMs; however, there are intrinsic limitations to obtaining highest resolution images due to lens aberrations, diffraction effects and the nature of sample/electron beam interaction. Although the instrument design has come a long way, especially for low voltage imaging, the problems associated with specimen/beam interactions and in particular observation of electron beam sensitive materials can still be insurmountable. This suggests that a different type of imaging technology, such as He Ion Microscope (HeIM), may satisfy the criteria of ultra-small spot size and allow imaging of beam sensitive materials with minimal damage that will be required to advance imaging of new materials at the nanoscale.

Helium Ion Microscope utilizes a scanning helium ion beam to image specimens. The design principle is based on field ion microscope operating in UHV environment with a cryogenically cooled sharp tungsten tip and to which He ions are being introduced (Figure 8.2). The uniqueness of the HeIM instrument lies in the shaping of the tip (Figure 8.2, inset). The tip is manufactured in such a way that it is truncated by a trimer of atoms. The emission from the trimer occurs at relatively low voltage and the He gas arriving at the tip can only interact with just a few atoms. Under normal operating condition the gun is centered in such a way that only a single atom emission is used for imaging. HeIM provides better resolution (sub 1 nm level), higher surface sensitivity due to substantially reduced interaction volume and completely new and different image contrast mechanisms that could provide information previously unattainable by low voltage SEM, especially in the

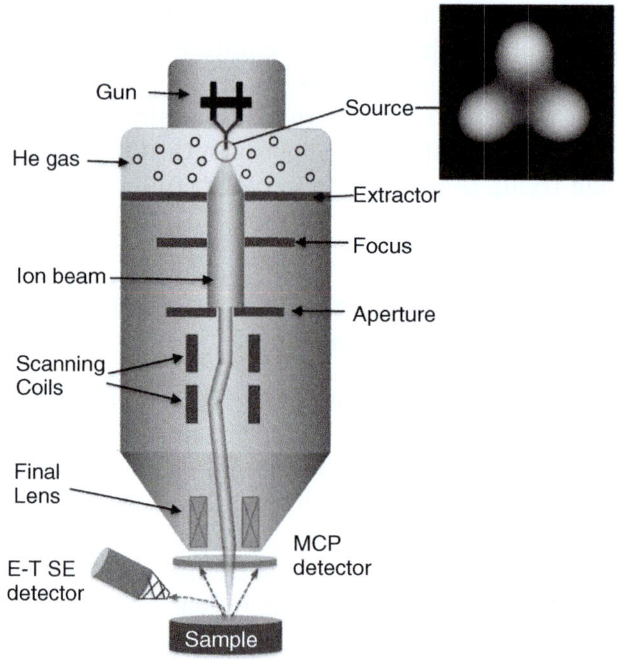

Figure 8.2 Simplified schematic of the Orion He ion microscope column, indicating source, apertures and ET SE and MCP detector configuration. (Inset) Image of W atomic trimer showing He ion emission from three atoms, one atom is selected by way of an aperture placement to produce the beam for imaging.

case of beam sensitive and non-conductive materials. The comparison of beam/specimen interaction in a He ion microscope to a traditional SEM and the resulting signal generation are shown in Figure 8.3(a). In the HeIM the interaction volume is intrinsically and significantly smaller than in an SEM at the corresponding accelerating voltage. The SE image produced in HeIM closely resembles a lower accelerating voltage condition in the SEM (Figure 8.3b). In the SEM and HeIM a smaller interaction volume results in better spatial resolution (Figure 8.3c), better surface-specific information and limit the contribution from back-scattered electrons and secondary electrons generated by back-scattered electrons/ions (SE_{II}) that tend to produce blurring of the image due to delocalization of signal. Furthermore, other effects can be expected while imaging with ions, such as ion implantation, displacement of atoms and removal of material (ion milling). These features make HeIM a fascinating and still barely explored alternative to ultra-low voltage

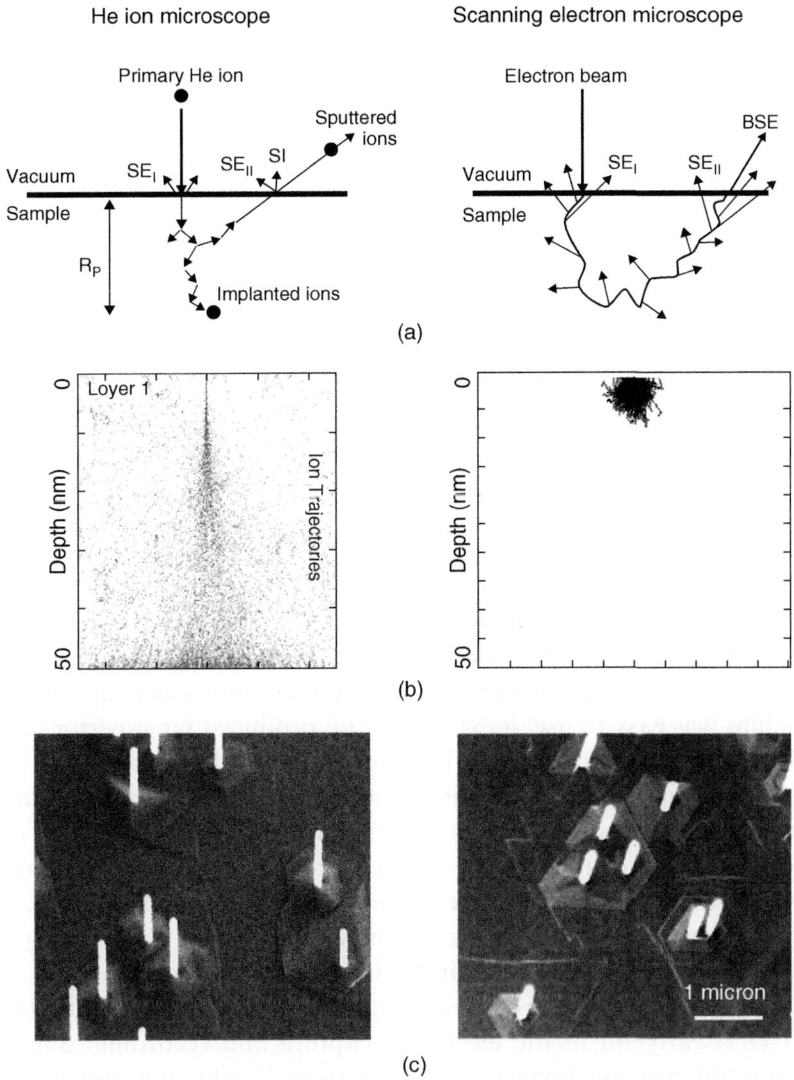

Figure 8.3 Comparison between HeIM and low-voltage SEM: (a) signal generation, including SE (secondary electrons), SI (secondary ions), BSE (back-scattered electrons); (b) TRIM simulation of ion/sample interaction using 20 kV He ion beam compared to EFS simulation of electron/sample interaction using 0.5 kV electron beam; (c) InAs nanowires imaged with HeIM at 20 kV and SEM at 0.5 kV.

imaging in SEM, with significantly better level of resolution than what can be achieved in the current state of the art scanning electron systems.

Moreover, HeIM can be also utilized as a fabrication tool with the ability to etch and modify materials with the bombardment of the helium ions, allowing the application of the instrument to lithography. It has been shown that direct etching is possible of materials raging from gold to graphene. A recent addition of *Scanning* highlighted the best of the key features and applications of HeIM with a hint at the possible directions in development (Bell *et al.*, 2012).

8.3 CAMERAS, DETECTORS, AND OTHER ACCESSORIES

8.3.1 The Direct Electron Detector

CCD (charge coupled device) cameras have made their way very successfully into multitude of areas of imaging sciences, partly driven by the consumer CCD camera industry and partly by the push by the astronomical sciences to develop remote automated telescope operation. The adoption by the electron microscopy community was painfully slow, since film was easy to use, inexpensive and produced far superior results to the early CCD cameras. Film was regarded as the gold standard for image recording for transmission electron microscopy. Manipulation of film was, of course, messy and cumbersome and perhaps the most appealing aspect of moving towards electronic imaging was to remove the use of messy chemicals in the EM laboratory environment. For automated and high throughput data acquisition in electron tomography and cryo-electron microscopy, a digital solution was needed.

Some researchers had proposed using Direct Electron Detectors (DED) or (DDD) early on in the electronic capture and recording. Since an electron microscope images with "electrons," why not use a direct electron detector plate much like it is used in particle physics to capture and record images (e.g., Bell, 1995). However, the cost of detectors at that time was extremely prohibitive. The economies of scale meant that adapting consumer technologies provided the fastest route to a workable CCD system.

The charge coupled device (CCD), familiar to all digital photographers, which provides digital readout, has become a popular image recording option for these applications (Fan and Ellisman, 2000). However, due to radiation damage and signal saturation (Roberts *et al.*,

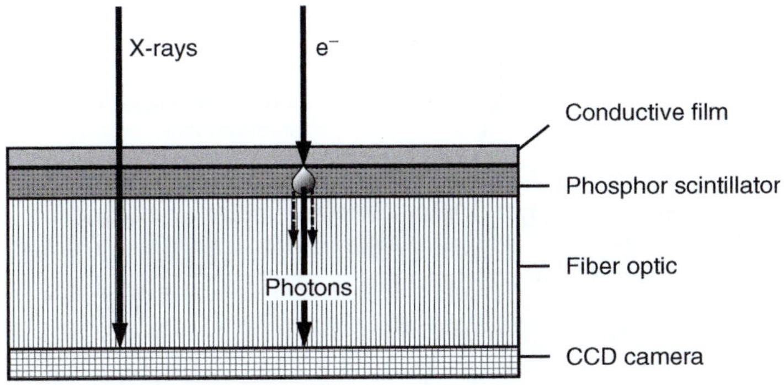

Figure 8.4 Fiber optic coupled CCD imaging system for TEM.

1982), CCDs cannot be placed in the direct path of the electron beam. A phosphorescent scintillation screen is required to convert the electron image to a photonic image that can be relayed to the CCD camera, either by fiber optics or a lens system. Unfortunately, the electron-to-photon conversion process in the scintillation screens generally limits the spatial resolution of CCD systems and more importantly for low voltage microscopy the ultimate sensitivity and collection efficiency of the camera (see Figure 8.4). Each incident electron will produce a footprint or scattering profile that is quite large: for 200 kV electrons, the footprint can have a full width at half maximum (FWHM) of about 50 μm; for 80 kV it can be nearly twice that size. CCD camera spatial resolution is limited by the size of this footprint. There are two main issues to be considered for imaging with low energy electrons: the first is the large fluctuations in the energy deposition; the second is the degradation of the point spread function (PSF) due to the $1/E$ increase of the electron multiple scattering in the detector itself.

In the recent years there has been a significant effort in development of DDD CMOS detectors, based on the active pixel sensor (APS) technology (Milazzo *et al.*, 2005; Xuong *et al.*, 2004). The active pixel sensor (APS) was originally developed to detect charged particles (Kleinfelder *et al.*, 2002, 2003, 2004; Matis *et al.*, 2002, 2003; Turchetta *et al.*, 2001). A prototype direct detection device (DDD) with 512×550 pixels and 5 μm pixel pitch was extensively tested and proven to be radiation tolerant for an accumulated dose of $>3.3 \times 10^6$ electrons/pixel, enough for several months of low-dose image acquisition (Jin *et al.*, 2007; Xuong *et al.*, 2007).

The APS utilized for the DDD comprises a sensitive region consisting of a particularly pure and lightly doped layer of silicon that is epitaxially

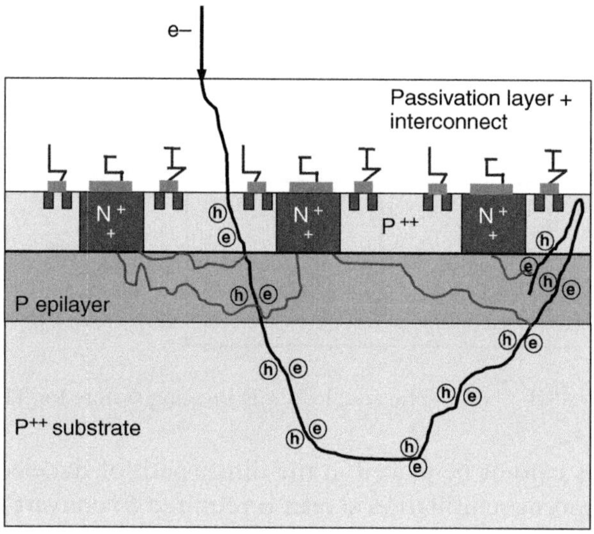

Figure 8.5 Schematic of APS pixel cross-section in Direct Electron Detector.

grown on top of a more heavily doped silicon wafer. Into this epitaxial silicon region, shallow, p-type wells are implanted, wherein n-type transistors can be fabricated. Inside each pixel is a diode formed by the interface between an n-well implantation and the p-epitaxial region, as well as three transistors that read out the charge collected on the diode.

High-energy electrons from the electron microscope leave trails of ionization within the silicon. These ionization electrons quickly recombine in the highly doped bulk silicon, but have a reasonably long lifetime in the high purity epitaxial region (Figure 8.5). The epitaxial silicon is sandwiched between two more heavily doped silicon regions (the bulk silicon wafer and the p-wells at the surface of the epitaxial silicon). This creates potential barriers to the electrons, causing liberated charge to be confined within the epitaxial silicon. With an 8 mm-thick epitaxial region, an incident electron of 200 or 300 kV will liberate about 1500 electrons in the epitaxial layer, which is significantly higher than the noise of the pixel's readout circuitry. The electrons are confined within the epitaxial region and diffuse randomly through it until they are collected by one of the pixel diodes. Each pixel effectively integrates the collected ionization electrons. At the end of the integration period, the contents of the sensor array are read out, digitized, displayed, and ultimately stored. Detector chips, including the prototypes, are fabricated in a standard digital CMOS process, as used for commercial microcomputers, and so on, making them convenient and inexpensive to fabricate. Unlike conventional digital cameras, in which pixels are sensitive to only a fraction of the light that strikes them, the "fill factor" (i.e., the proportion of each pixel area that is sensitive to incident electrons) is 100%, since the sensitive epitaxial layer is continuous beneath the readout circuitry.

With the use of specimen drift correction and large area mosaic techniques, the DDD can increase the utility of electron tomography for studying large-scale biological structures. The high sensitivity feature would also make DDD suitable for many low-dose imaging applications such as cryo- EM and cryo-electron tomography.

Despite concerns about increased noise in the low frequency range due to backscattered electrons (Milazzo *et al.*, 2010), there is strong evidence that the improved performance of the DDD results will translate into higher quality 3D density maps of biological specimens derived from low-dose imaging and single particle reconstruction methods (Glaeser *et al.*, 2011). The new detectors provide advantages over currently used CCD cameras in both the spatial resolution and the signal-to-noise ratio of the acquired images. The DDD detector achieves signal to noise ratios equivalent to CCD cameras at a fraction of the electron dose with improved quantum efficiency, capturing more information from a given electron dose, and accelerating the rate at which the signal-to-noise ratio improves over the exposure period.

8.3.2 Silicon Drift Detectors for Low kV Nanoanalysis

Recent advances in Silicon Drift Detector (SDD) design means that the latest generation of these EDS detectors not only offer all the advantages of being liquid nitrogen-free and excellent energy resolution at high count rates, but now large active area detectors allow large amounts of data to be collected in shorter time periods, at lower excitation voltages.

The Energy Dispersive X-ray (EDS) detector converts the energy of each individual X-ray into a voltage signal of proportional size (Garratt-Reed and Bell, 2003). This is achieved through a three-stage process. Firstly the X-ray is converted into a charge by the ionization of atoms in the semiconductor crystal. Secondly this charge is converted into the voltage signal by the FET preamplifier. Finally the voltage signal is input into the pulse processor for measurement. The output from the preamplifier is a voltage "ramp" where each X-ray appears as a voltage step on the ramp. The EDS detectors are designed to convert the X-ray energy into the voltage signal as accurately as possible. At the same

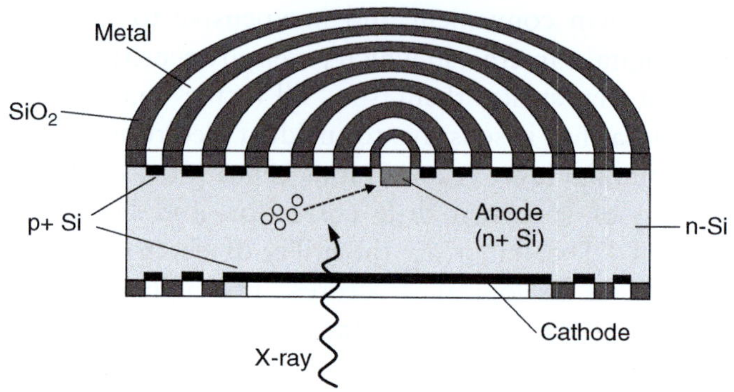

Figure 8.6 Silicon drift detector chip cross-section.

time electronic noise must be minimized to allow detection of the lowest X-ray energies.

The silicon drift detector (SDD) sensor is fabricated from high purity silicon with a large area contact on the entrance side facing the incoming X-rays. On the opposite side there is a central, small anode contact, which is surrounded by a number of concentric drift electrodes (Figure 8.6). When a bias is applied to the SDD detector chip and the detector is exposed to X-rays, it converts each X-ray detected into an electron cloud with a charge that is proportional to the characteristic energy of that X-ray. These electrons are raised into the conduction band of the silicon semiconductor and leave behind holes that behave like free positive charges within the sensor. The electrons are then "drifted" down a field gradient applied between the drift rings to be collected at the anode.

SDD detectors clearly possess an important advantage because they maintain best resolution performance at much higher count rates than for Si(Li) detectors. Higher productivity is achieved by collecting data faster with no loss of analytical performance. However, of course many more X-rays must be generated and collected for the productivity advantages of SDD technology to be realized. While this is possible for some applications on stable samples, it is undesirable, difficult or impossible for beam sensitive, unstable, easily contaminated samples, or for nanoanalysis at low kV where much fewer counts are generated.

The solution to extend the benefits of SDD to a wide range of applications is to fabricate larger detectors that can collect more X-rays under the same SEM operating conditions. When Si(Li) detectors are made with larger areas, as the area increases, the anode capacitance and noise increases. As a result, with Si(Li) detectors, 30 mm^2 has so far proved to be the about the limit . However, in SDD, instead of the whole back surface of the sensor being the anode (as it is with Si(Li)), the drift field brings the electrons to a very small anode, thus drastically reducing readout capacitance. Furthermore, the SDD anode capacitance is not affected by the area, so the design offers the potential for much larger sensors that offer good resolution.

The longer the process time (TP), the lower the voltage noise. Provided the leakage current noise is still low, the resolution of the peak displayed in the spectrum is improved, and it becomes easier to separate or resolve, from another peak that is close in energy. However, there is a trade-off between the process time that is used, and the speed at which data can be measured. The longer the process time, the more time is spent measuring each X-ray, and the fewer events that can be measured. The longest process time used by a processor gives the best resolution possible, while the shortest process time gives the maximum throughput into the spectrum, but there will be some loss of resolution.

The drawback with large area SDDs is the increased leakage current and the increased distance required to drift the signal charge cloud towards the readout anode. The leakage current increase can be nullified by increased cooling and the drift distance is minimized by use of a central anode. A central anode arrangement is more challenging with an integrated FET that sits inside the anode ring due to the need to protect the integrated FET from X-ray bombardment and avoid charge losses for X-rays that fall at the center of the device on top of the FET. These issues are avoided by using an external FET arrangement where the SDD anode is simply a small central electrode and sensors up to 100 mm^2 in size are available that offer identical resolution and analytical performance to equivalent 10–30 mm^2 sensors. A practical limit to detector size is the diameter of the EDS detector tube that goes through a port on the microscope column. In addition, the design of the magnetic electron trap becomes more challenging when the open area is larger. Although in principle a smaller detector placed closer to the specimen can provide equivalent solid angle, at closer distances there is more risk of damage and interference with microscope imaging performance.

Ballistic deficit in SDDs caused by larger area detectors would be an issue that becomes apparent for a low kV systems. The electron cloud generated by the incoming X-rays spreads out before it reaches the anode. This is a problem for SDD because of the relatively slow speed at which the electron clouds are drifted to the anode for signal collection. The longer the path distance between the point where the

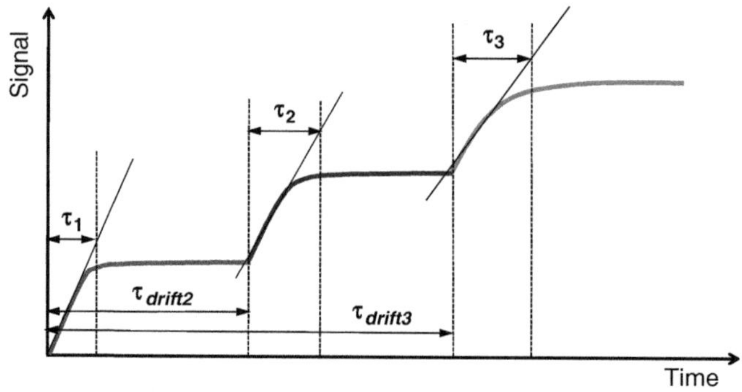

Figure 8.7 Electron drift distance and rise time in an SDD detector.

X-ray is absorbed and the anode, the more the electron cloud spreads out. When the cloud gets close to the anode it begins to induce an output signal but the full signal step is not achieved until the whole cloud is collected; the larger the cloud, the longer the time. As a result, the rise time of the signal pulse depends on the original point of interaction in the detector (see Figure 8.7).

To achieve very high count rates, the signals have to be measured fast using a short processing time. If too short a process time is used, the pulse may not reach its full peak height so there is a deficit in the measurement referred to as "ballistic deficit." Ballistic deficit therefore becomes an important issue as the process time decreases, the mode of operation for SDD at high count rate. Because the pulses from a large area detector will have a mixture of rise times, the resulting effect will be similar to incomplete charge collection, with tailing on the low energy side of peaks. The distortion of peak shape away from the ideal model will reduce the accuracy of spectrum processing, particularly for overlapping peaks. Therefore, to run large area SDD detectors at high count rates the issue of ballistic deficit must be accounted for. The problem is reduced by limiting the path distance of drift to the anode by using a symmetrical design. It is also possible to use special pulse processing techniques to make sure the measurement is insensitive to variations in rise time. Although this limits the maximum speed it does ensure that good analytical performance is maintained.

8.4 CONCLUSIONS

With advances in electron gun design, aberration corrected optics and improvements in the sensitivity of recording images and obtaining analytical information, the future for low voltage electron microscopy is very "bright". For example, the dedicated low voltage TEMs have showed that there is a market for the simple low voltage instrument for basic research, but the higher resolution aberration corrected instruments have certainly shown the way in what is possible by pushing the limits of performance.

Better detectors both for imaging and analysis are critical, making each electron in the low kV range count and obtaining useful information from every electron-sample interaction in the most efficient manner will be the focus on the future. For example direct electron detection enables the acquisition of low-noise images of delicate, biological samples and beam-sensitive materials that require low electron dose interactions to prevent radiation damage of the material. These detector improvements both for TEM and SEM will further push the limits of low voltage electron microscopy.

REFERENCES

M. Bendayan, I. Londono, and E. Paransky, Compartmentalization of pancreatic secretory zymogen granules as revealed by low-voltage transmission electron microscopy, *J. Histochem. Cytochem.*, **59**, 899–907 (2011).

D.C. Bell, Ph.D. thesis, University of Melbourne, Australia (1995).

D.C. Bell (ed.), Helium Ion Microscopy (special edition), *Scanning*, **34** (4), (2012).

T.H.P. Chang, D.P. Kern, and L.P. Muray, Microminiaturization of electron optical systems, *J. Vac. Sci. Technol. B*, **8**, 1698–1705 (1990).

E. Coufalova and A. Delong, Low voltage electron microscope I.—applications, *EUREM 12*, I183 (2000).

A. Delong, Low voltage TEM, *EUREM* **92**, 1, 79–81 (1992).

A. Delong, K. Hladil, and V. Kolařík, A low voltage transmission electron microscope, *Microscopy & Analysis*, **27** (Europe), 13–15 (1994).

A. Delong, K. Hladil, V. Kolařík, and P. Pavelka, Low voltage electron microscope I.—design, *EUREM 12*, Brno, Czech Republic, I197 (2000).

L.F. Drummy, J. Yang, and D.C. Martin, Low-voltage electron microscopy of polymer and organic molecular thin films, *Ultramicroscopy*, **99**, 247–256 (2004).

G.Y. Fan and M.H. Ellisman, Digital imaging in transmission electron microscopy, *J. Microsc.*, **200**, 1–13 (2000).

A.J. Garratt-Reed and D.C. Bell, *Energy Dispersive X-Ray Analysis in the Electron Microscope* (Microscopy Handbooks), Garland Science (2003).

R.M. Glaeser, G. McMullan, A.R. Faruqi, and R. Henderson, Images of paraffin monolayer crystals with perfect contrast: minimization of beam-induced specimen motion, *Ultramicroscopy*, **111**, 90–100 (2011).

L. Jin, A.C. Milazzo, S. Kleinfelder, *et al.*, The intermediate size direct detection detector for electron microscopy, *Proc. SPIE Int. Soc. Opt. Eng.*, **6501**, 65010A (2007).

S. Kleinfelder, H. Bichsel, F. Bieser, *et al.*, Integrated X-ray and charged particle active pixel CMOS sensor arrays using an epitaxial silicon-sensitive region, *Proc. SPIE*, **4784**, 208–217 (2002).

S. Kleinfelder, F. Bieser, Y.D. Chen, *et al.*, Novel integrated CMOS pixel structures for vertex detectors, *Nucl. Sci. Symp. Conf. Rec. IEEE 2003*, **1**, 335–339 (2003).

S. Kleinfelder, F. Bieser, Y.D. Chen, *et al.*, Novel integrated CMOS sensor circuits, *IEEE Trans. Nucl. Sci.*, **51**, 2328–2336 (2004).

F. Lednický, E. Coufalová, J. Hromádková, *et al.*, Low voltage TEM imaging of polymer blends, *Polymer*, **41**, 4909–4914 (2000).

H.S. Matis, F. Bieser, S. Kleinfelder, *et al.*, A CMOS active pixel sensor for charged particle detection, *Nucl. Sci. Symp. Conf. Rec. IEEE 2002*, **1**, 259–263 (2002).

H.S. Matis, F. Bieser, S. Kleinfelder, *et al.*, Charged particle detection using a CMOS active pixel sensor, *IEEE Trans. Nucl. Sci.*, **50**, 1020–1025 (2003).

A.C. Milazzo, P. Leblanc, F. Duttweiler, *et al.*, Active pixel sensor array as a detector for electron microscopy, *Ultramicroscopy*, **104**, 152–159 (2005).

A.C. Milazzo, G. Moldovan, J. Lanman, *et al.*, Characterization of a direct detection device imaging camera for transmission electron microscopy, *Ultramicroscopy*, **110**, 744–747 (2010).

L.P. Muray, C.A. Silver, and J.P. Spallas, Sub-100 nm lithography with miniature electron beam columns, *J. Vac. Sci. Technol. B*, **24**, 2945–2950 (2006).

L.P. Murray, Developments in low-voltage microscopy instrumentation, *Scanning*, **33**, 155–161 (2011).

P.T.E. Roberts, J.N. Chapman, and A.M. MacLeod, A CCD-based image recording system for the CTEM, *Ultramicroscopy*, **8**, 385–396 (1982).

W.P. Sharp, E. Coufalova, A. Delong, *et al.*, Use of the low voltage TEM with biological specimens: a feasibility study, *Microscopy and Microanalysis*, Philadelphia (2000).

J.P. Spallas, C.S. Silver, L.P. Muray, *et al.*, A manufacturable miniature electron beam column, *Microelectron Eng.*, **83**, 984–989 (2006).

R. Turchetta, J.D. Berst, B. Casadei, *et al.*, A monolithic active pixel sensor for charged particle tracking and imaging using standard VLSI CMOS technology, *Nucl. Instrum. Meth. Phys. Res.*, A **458**, 677–689 (2001).

N.-H. Xuong, A.-C. Milazzo, P. LeBlanc, *et al.*, First use of a high-sensitivity active pixel sensor array as a detector for electron microscopy, *Proceedings SPIE*, **5301**, 242–249 (2004)

N.H. Xuong, L. Jin, S. Kleinfelder, *et al.*, Future directions for camera systems in electron microscopy, *Meth. Cell Biol.*, **79**, 721–739 (2007).

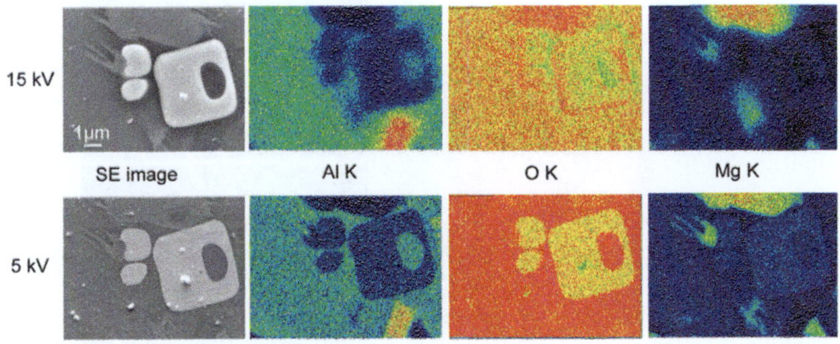

Figure 1.18 Advantage of X-ray microanalysis at low voltages in SEM. WDS maps of polished section of ore from Miake Island. Elemental maps of Al, O and Mg shown alongside corresponding SE images, taken at 15 kV (top) and 5 kV (bottom).

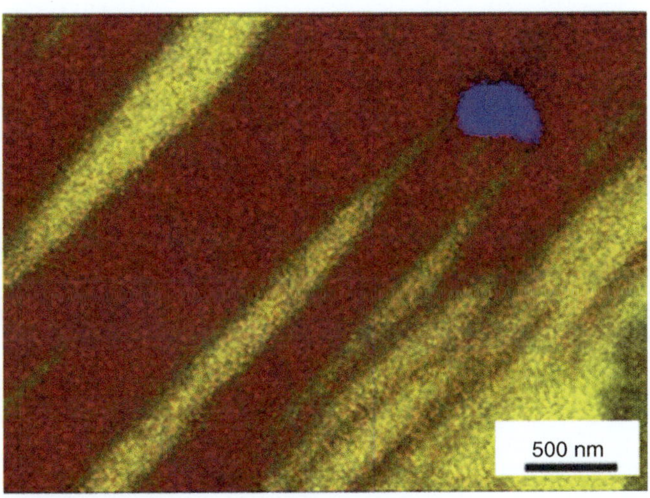

Figure 2.5 EDS map of Sphene $CaTi(Si0_4)(O, OH, F)$ (yellow), Apatite (blue) in a magnesium aluminosilicate matrix (red) with sub-100nm resolution. The map was collected at 8 kV for 20 min using beam current of 20 nA.

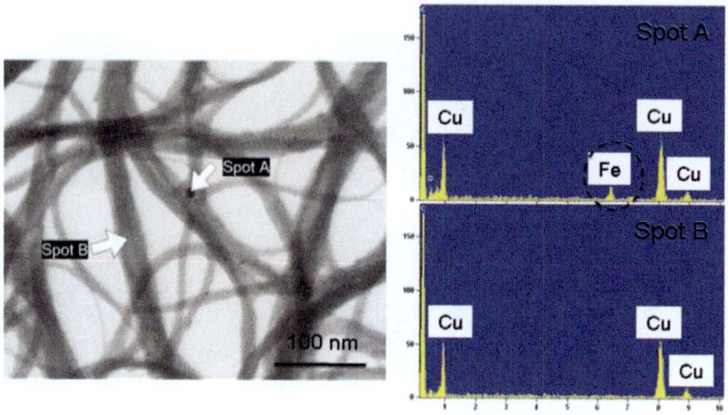

Figure 4.5 Left: image of carbon nanotubes in STEM mode taken with JSM-7600F at 30 kV. Right: spot EDS analyses of two different locations: catalyst particle (Spot A), and CNT matrix (Spot B).

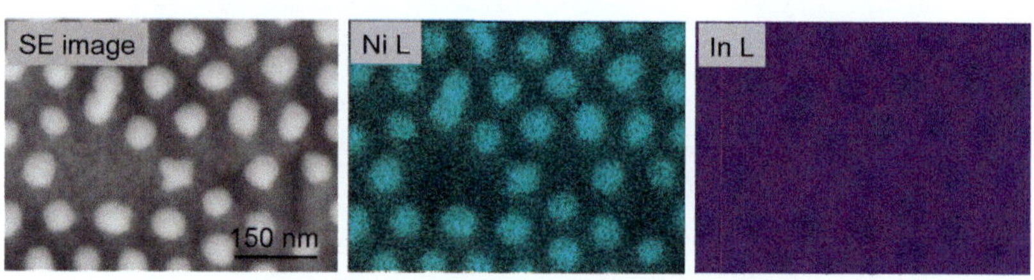

Figure 4.15 Ni pillars on ITO (indium tin oxide). 3 kV EDS mapping (30 mm^2 Oxford SDD), 5 min acquisition with drift correction.

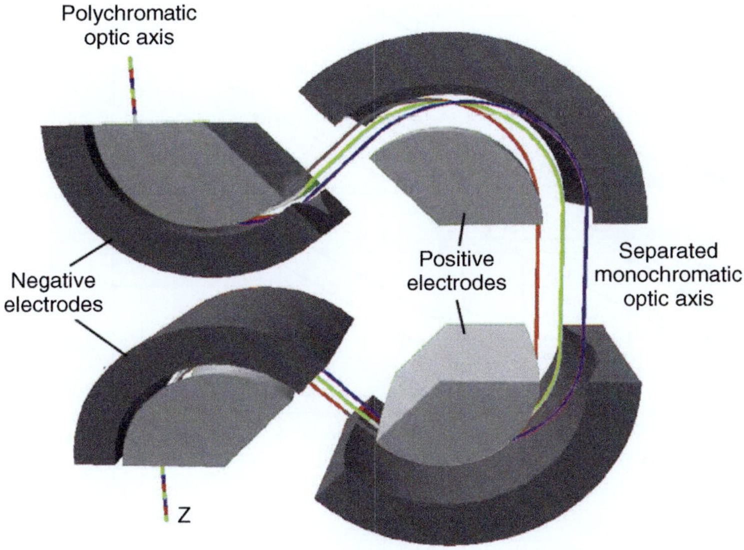

Figure 5.3 The Omega electron monochromator (CEOS) (Reproduced with permission from Bell et al., 2012).

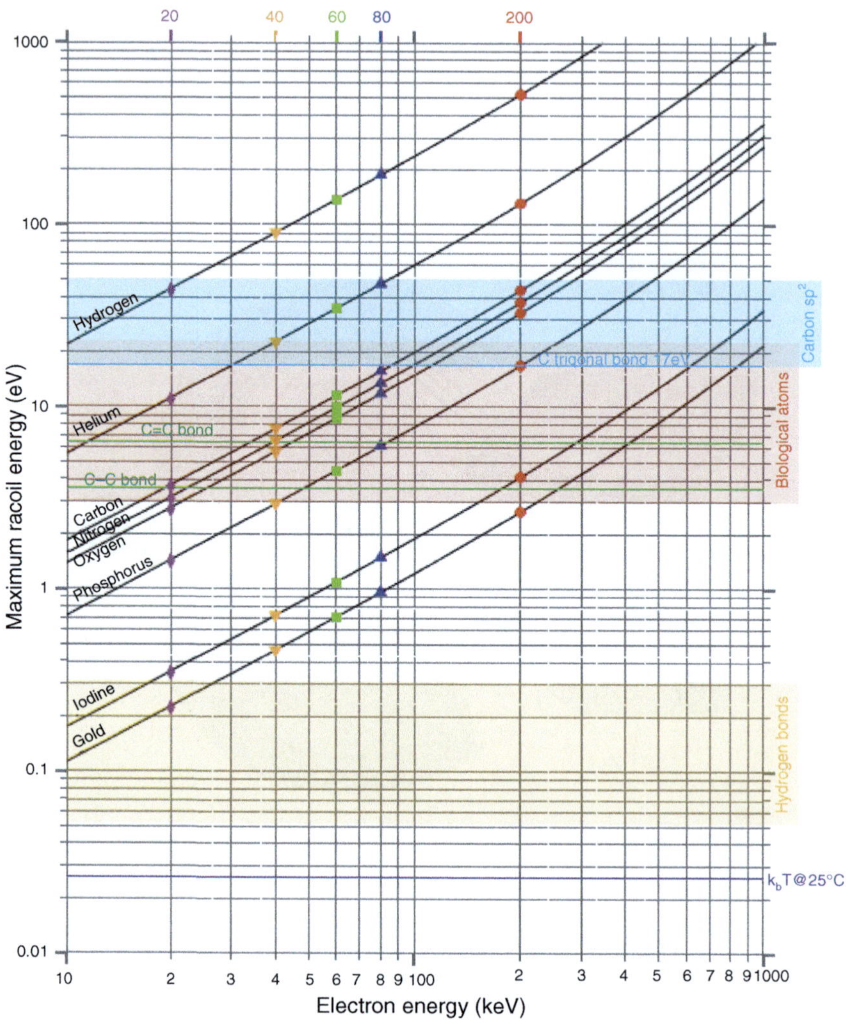

Figure 5.4 Maximum atomic recoil energy as a function of incident electron energy, colored regions indicate the bond breaking energies. (Reproduced with permission from Bell *et al.*, 2012).

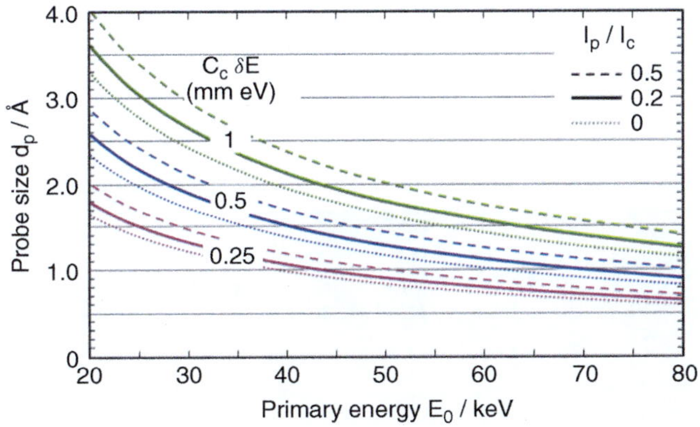

Figure 6.1 The dependence of the probe size d_p on the primary energy for three values of the chromatic product $C_c\delta E$ and three different ratios I_p/I_c of the actual beam current to the coherent current.

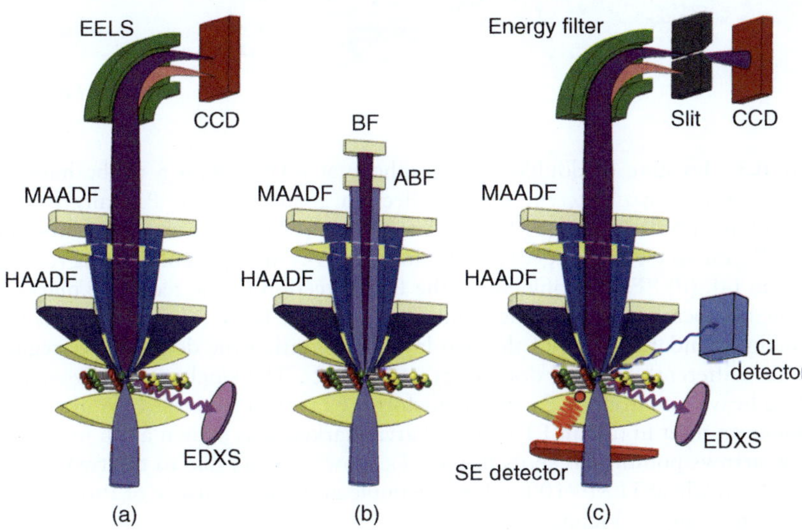

Figure 6.2 Different ways of configuring detectors in a STEM in which the electrons travel upwards. (a) "Standard" materials science set-up combining ADF imaging (either HAADF or MAADF, or both) with EELS and optionally also EDXS. (b) Set-up for imaging light elements using annual bright field (ABF) and on-axis bright field (BF) detectors. (c) The "physicist's set-up" for capturing and analyzing 2D, energy-filtered diffraction patterns plus several other analytical signals.

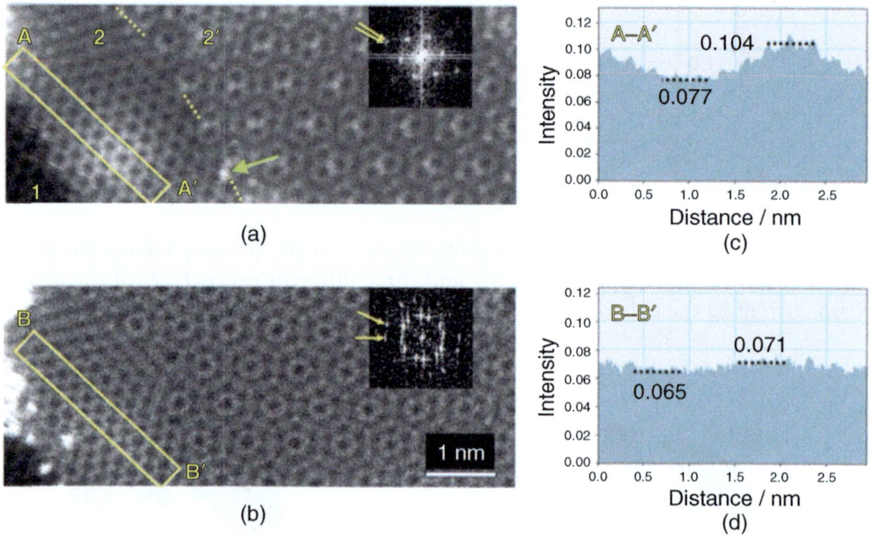

Figure 6.3 Imaging of double-layer graphene at 60 keV with a probe half-angle of 30 mrad and two different signal collection geometries. (a) Partially coherent (37 mrad inner half-angle). (b) Largely incoherent (72 mrad inner ADF half-angle). (c) "Fat" (transversely averaged) profile along the A-A' direction of the long rectangle shown in (a). (d) "Fat" profile along the B-B' direction of the rectangle marked in (b). The numbers in (a) mark the number of graphene layers, the dotted lines mark a domain boundary in one of the two layers, separating the double-layer region 2 from the differently-rotated double layer region 2′. The single green arrow in (a) marks a heavier adatom. A Fourier Transform (FT) taken from the area marked 2 is shown as an inset in (a), an FT from the area marked 2′ is shown as an inset in (b). Yellow arrows pointing to spots in the FTs show maxima due to the two separate layers. Nion UltraSTEM™100, 60 keV. Sample and data courtesy of Professor J.C. Meyer, University of Vienna.

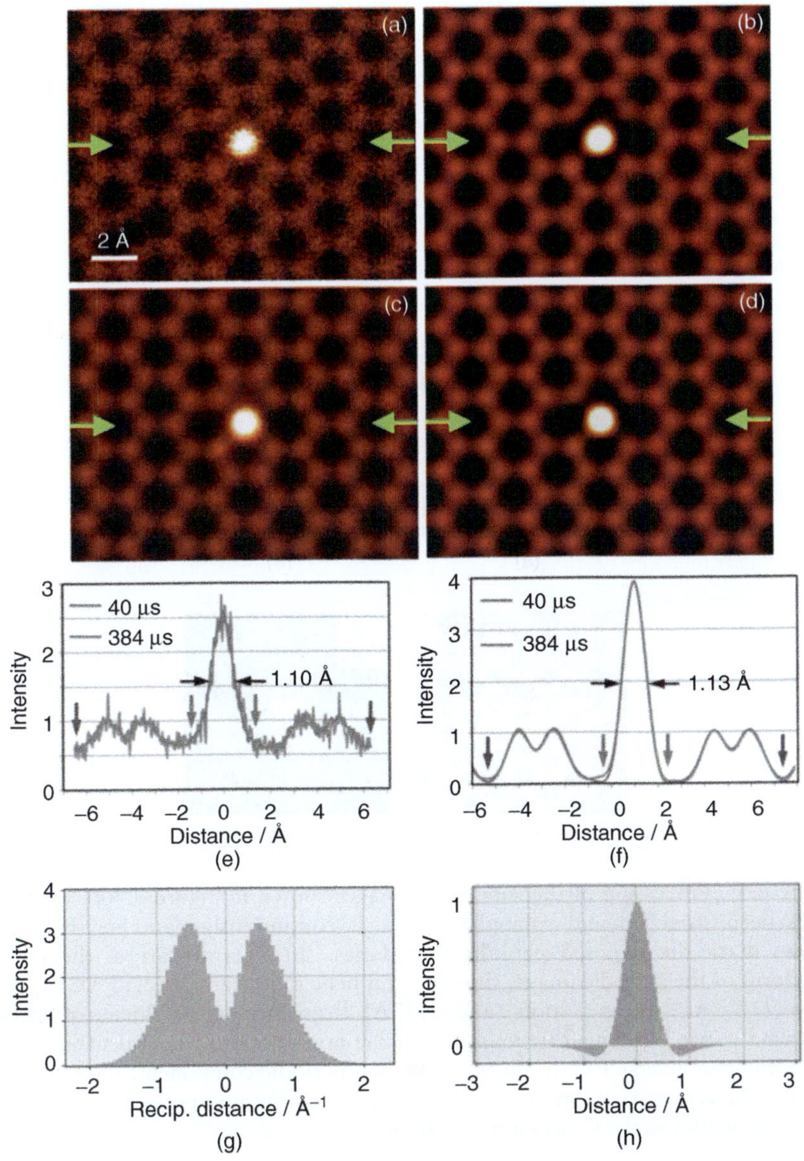

Figure 6.4 MAADF images of a Si atom embedded in graphene taken in two different ways, processed versions of the images, and line profiles through the images. (a) Single MAADF image taken with 40 μs/pixel acquisition time. (b) Image (a) processed with a smoothing and probe skirt-removing filter. (c) Sum of 16 images taken with 24 μs/pixel. (d) Image (c) processed with a smoothing and probe skirt-removing filter. (e) Line profiles taken between the two green arrows in images (a) and (c). (f) Line profiles taken between the two green arrows in images (b) and (d). (g) The smoothing and skirt-subtracting filter as it was used in the FT domain. (h) Real-space shape of the filter. Blue arrows in (e) and (f) mark the centers of graphene rings, red arrows mark ±1.42 Å distance from the center of the Si atom, that is, the C-C nearest neighbor separation. UltraSTEM™ 100, 60 keV.

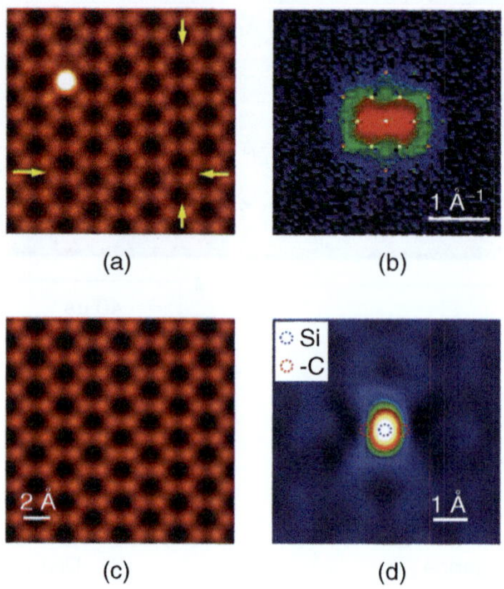

Figure 6.5 Processing of the sum of 16 images shown in Figure 6.4(c). (a) Image that's been stretched and rotated to make it continuously periodic; the yellow arrows mark the imposed periodic boundaries. (b) Central part of the Fourier transform of the periodic image, showing graphene reflections at $(2.13\,\text{Å})^{-1}$ (yellow spots), $(1.23\,\text{Å})^{-1}$ (orange spots), and $(1.07\,\text{Å})^{-1}$ (green spots). (c) Image of "perfect graphene lattice" obtained by masking off the non-periodic component in (b) and performing an inverse FT. (d) Image of the non-periodic component, showing the Si atom at the center, two "minus" carbon atoms, and the non-periodic variations of image intensities at the C sites in the graphene.

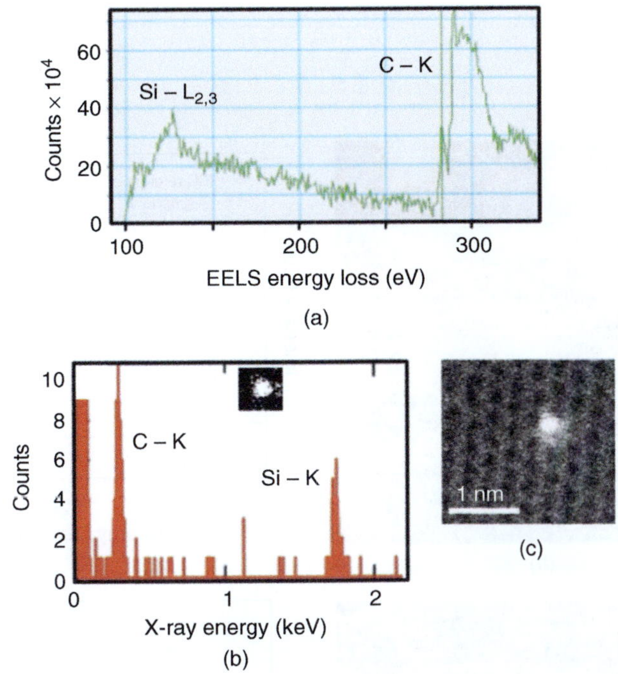

Figure 6.6 EEL and EDX spectra collected from a single Si atom. (a) EEL spectrum collected from a Si atom embedded in monolayer graphene acquired while the atom was being tracked using a 6×6 Å tracking window. (b) EDX spectrum collected simultaneously. (c) ADF survey image of the Si atom acquired just after the spectra were collected, with 2 s acquisition time. The small image insert in (b) shows one of the many tracking images acquired in 60 ms each. $I_p = 190$ pA, total acquisition time = 224 s, time the probe spent directly over the atom ~10 s. Nion UltraSTEM™100, Gatan Enfina EELS, Bruker XFlash 5030T EDX spectrometer, 60 keV.

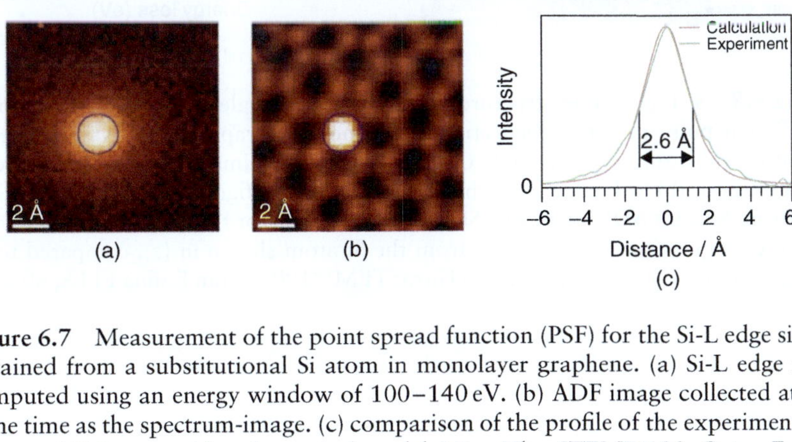

Figure 6.7 Measurement of the point spread function (PSF) for the Si-L edge signal obtained from a substitutional Si atom in monolayer graphene. (a) Si-L edge map computed using an energy window of 100–140 eV. (b) ADF image collected at the same time as the spectrum-image. (c) comparison of the profile of the experimentally measured Si-L map with a theoretical model. Nion UltraSTEM™100, Gatan Enfina EELS, 60 keV.

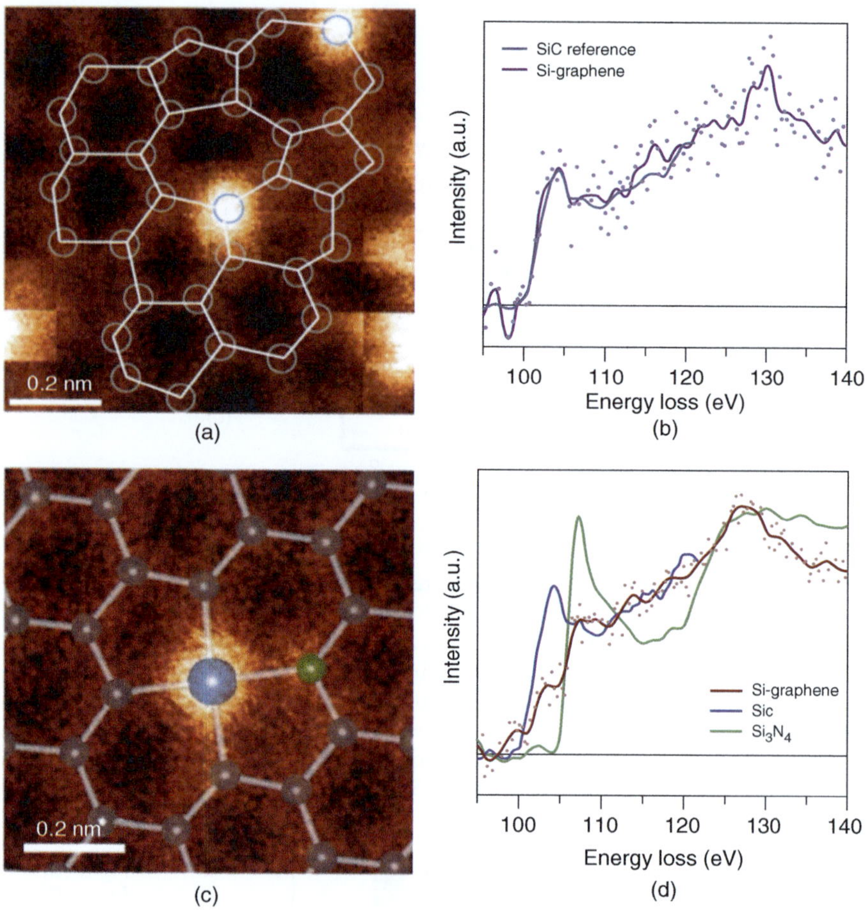

Figure 6.8 Si L-edge fine structures obtained from single substitutional Si atoms in different point defect configurations in monolayer graphene. (a) ADF image of a Si atom substituting for a single C atom (in center of image), in modified type A configuration. (b) Si-L edge from the Si atom shown in figure (a) compared to Si-L edge from SiC. (c) ADF image of a Si atom substituting for two C atoms, in modified type B configuration. (d) Si-L edge from the Si atom shown in (c), compared to Si L-edges from SiC and β-Si_3N_4. Nion UltraSTEM™100, Gatan Enfina EELS, 60 keV.

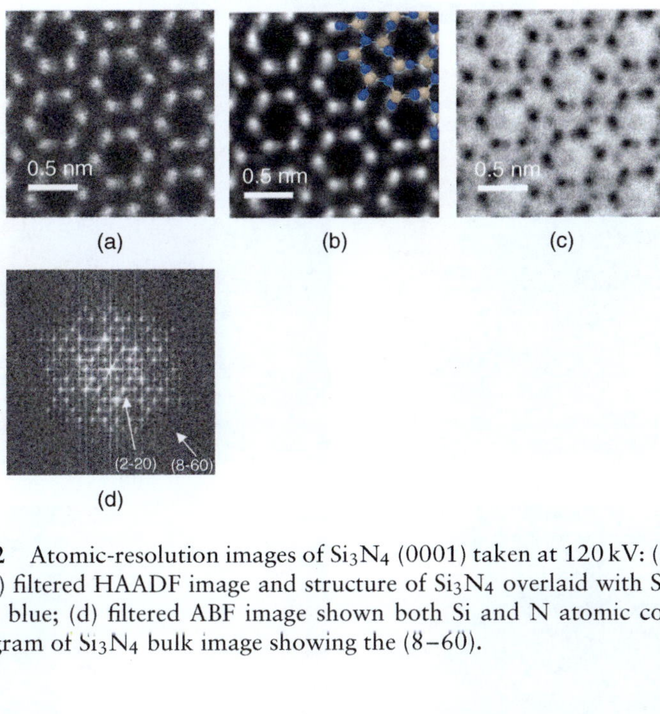

Figure 7.2 Atomic-resolution images of Si_3N_4 (0001) taken at 120 kV: (a) HAADF image; (b) filtered HAADF image and structure of Si_3N_4 overlaid with Si in brown and N in blue; (d) filtered ABF image shown both Si and N atomic columns; (d) diffractogram of Si_3N_4 bulk image showing the (8–60).

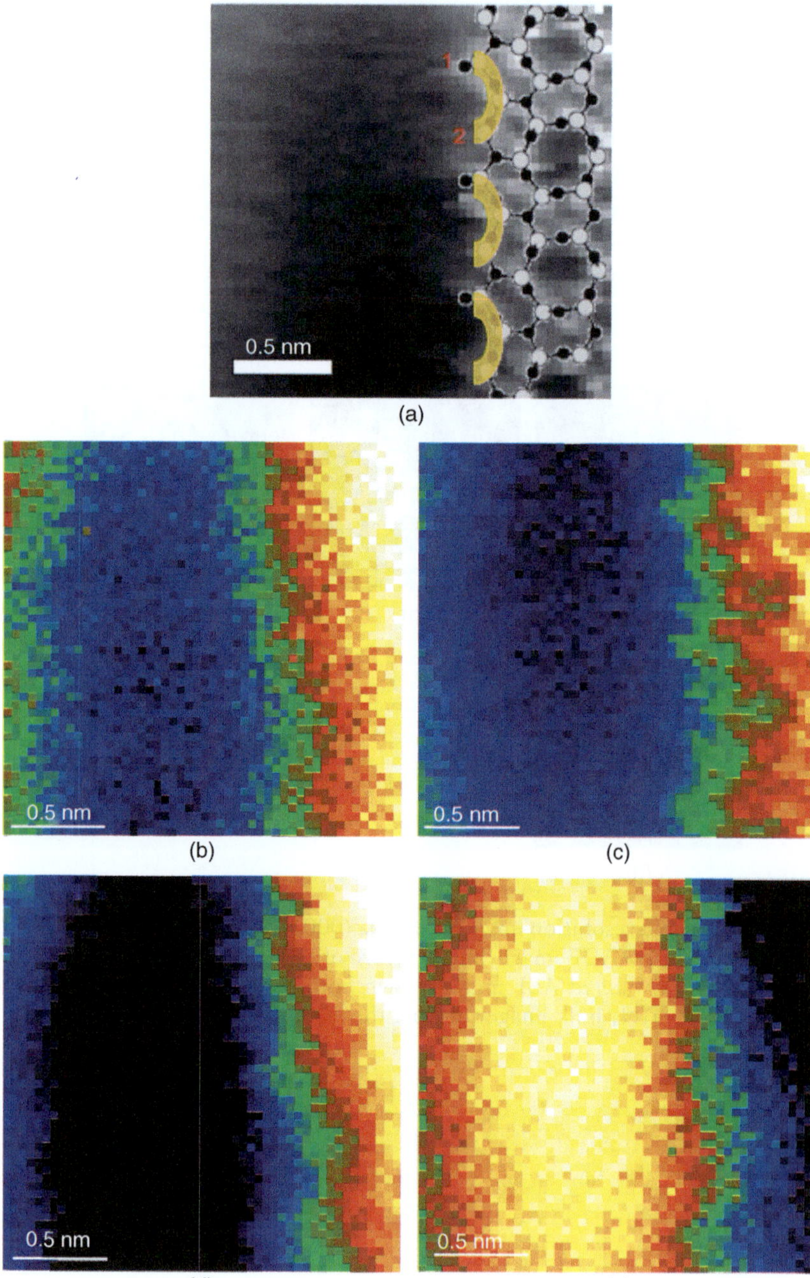

Figure 7.6 Spectrum imaging of Si_3N_4/SiO_2 interface at the atomic scale: (a) HAADF signal acquired simultaneously with the spectrum image shown in (b–e); (b) integrated Si L-edge intensity; (c) integrated extended L-edge fine-structure (200 eV above the edge onset); (d) N K-edge integrated intensity; (e) O K-edge integrated intensity.

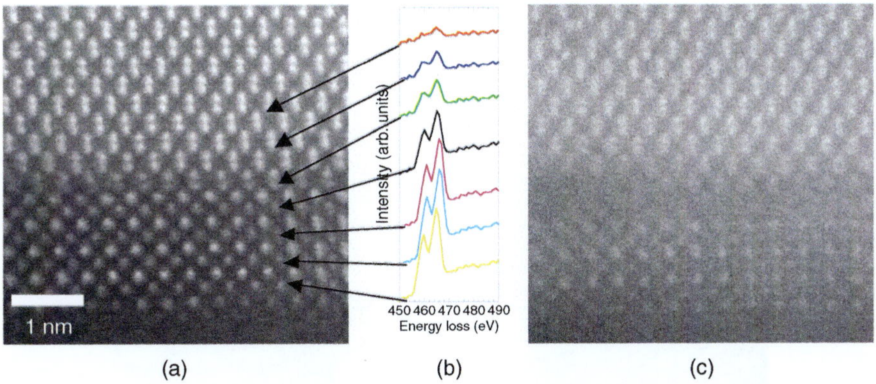

Figure 7.7 Atomic-resolution *Z*-contrast image of $SrTiO_3$/GaAs. (a) After a single image scan across the field of view. (b) Ti L-edge taken from the position indicated in (a), showing clear signs of Ti diffusion into the top-layers of the GaAs support. (c) Same region as shown in (a) after EELS measurements. The interfacial structure and the $SrTiO_3$ show signs of severe beam damage. The electron energy for this experiment was 100 keV.

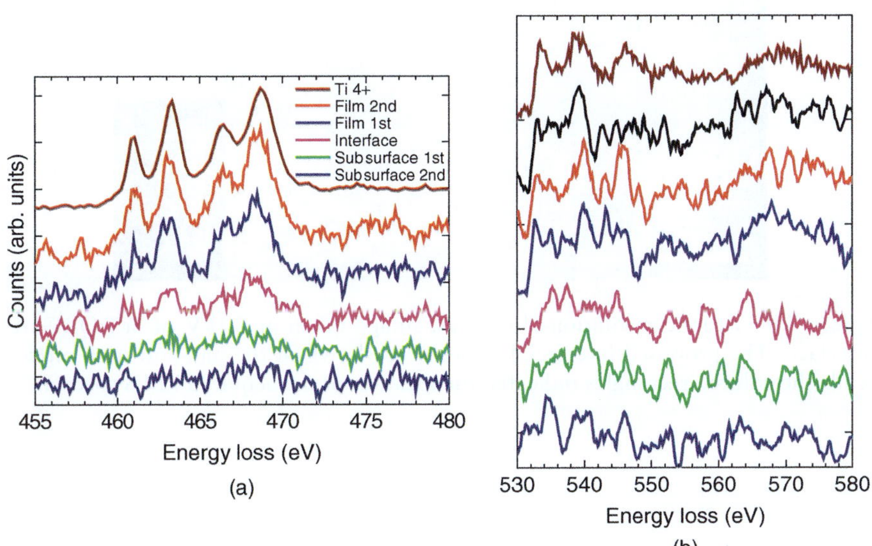

Figure 7.8 (a) Titanium L-edge and (b) oxygen K-edges across the $SrTiO_3$/GaAs interface. The location of the spectra is indicated by the colored rectangle in Figure 7.8.

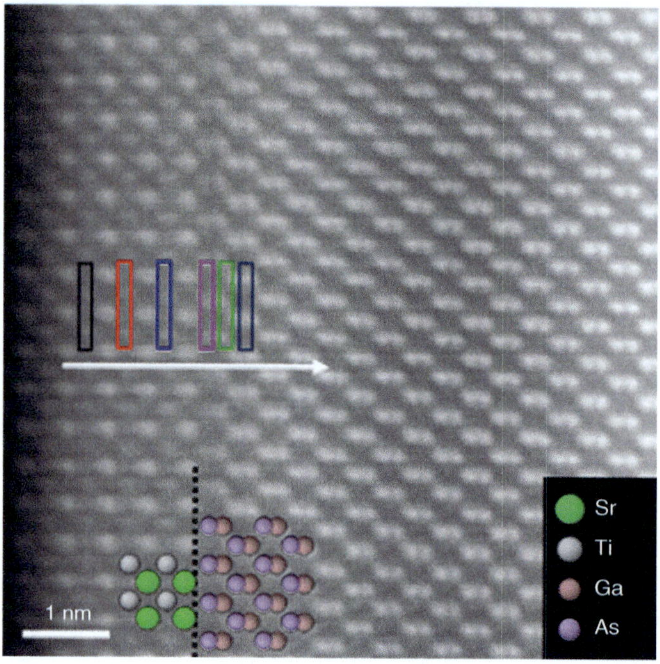

Figure 7.9 Atomic resolution *Z*-contrast image taken at 80 kV of $SrTiO_3/GaAs$ interface. The proposed interfacial structure is shown in the ball-and-stick model, while the colored rectangles indicated the location of the acquired EELS spectra.

Index

Aberrations
Chromatic, 14, 27, 32, 37, 40, 41, 50, 58, 64, 67, 74, 97, 100, 103, 121, 123, 126, 186
Spherical, 13, 27, 40, 41, 50, 58, 61, 74, 98, 104, 109
Diffraction, 15, 32, 58, 60
Correction (SEM), 41, 50, 58, 67
Correctors, 99, 103, 110, 164
Diffraction limit, 122
Airy disc, 140
Aperture angle control lens (ACL), 35, 38, 51, 91

Beam deceleration, 37, 41, 65, 81, 87
Biological specimen imaging, 87
Brain (thin sections), 89
Pollen, 87
Bacteriophage, 88
Biological specimen imaging (TEM), 101, 106
Boersch effect, 39
Boron nitride, 113, 120, 136
Brightness, 124

Cameras
CCD, 186, 192
Direct electron detector (DDD or DED), 192
Cathode Lens, 41
Cathodoluminescence, 92
Charge balance (specimen charging), 74, 81, 87
Convergence angle, 13, 15, 32, 38, 58, 163

Delocalization, 149
Detectors (SEM)
Secondary, 43, 66, 78, 85
Backscatter, 45, 66, 76, 79, 88
LABe, 46, 76
EsB, 46
DBS, 46
STEM-in-SEM, 79, 82, 87, 89
Detectors (STEM)
High angle annular dark field detector (HAADF), 164, 166, 173,
Annular bright field detector (ABF), 166, 168, 171, 173
Diffraction pattern, 131, 133
Dual Beam (FIB), 63

Elastic scattering, 8, 18, 27
EDS, 75, 80, 91
EELS (electron energy loss spectroscopy), 99, 164, 168, 173, 175
Electron detectors (STEM)
Annular bright field, 119, 129, 154
High annular dark field (HAADF), 129, 131

Low Voltage Electron Microscopy: Principles and Applications, First Edition.
Edited by David C. Bell and Natasha Erdman.

Electron detectors (STEM) (*continued*)
  Medium angle annular dark field (MAADF), 136, 138, 140
  Secondary, 121
Electron dose, 101, 109, 120, 151, 154
Electron source
  Brightness, 32, 61, 101, 105, 110
  Thermionic, 33
  LaB_6, 33
  Schottky (thermal) FE, 33, 59, 98, 101, 109
  Cold cathode FE, 34, 59, 97, 120, 124, 164, 168
  In-lens FE-SEM gun, 35
  Energy spread, 58, 59
Energy filtering (SEM), 43
  EsB, 46
  ExB, 43
  r-filter, 43, 77
Envelope function, 104, 110
Energy spread, 97, 103, 110, 113, 122
Environmental stability, 32, 64, 164

Graphene, 83, 112, 120, 127, 132, 135, 165

Helium ion microscope, 189

Image filtering , 139, 143
Immuno-labeling, 76, 87
Ionization (inner shell), 146, 149

LEEM (LEED), 67, 99
Lenses (SEM)
  Electromagnetic, 36
  Hybrid or compound, 36, 41
  Immersion, 37, 43
  In-lens SEM, 37
Low-dose imaging, 65, 69

Monochromator, 3, 14, 27, 36, 57, 59, 103, 186
Monte-Carlo simulation, 5
Multislice simulation, 112

Nanoparticles, 76, 110
Nanoporous materials, 81
Nanorods, 109
Nanotechnology, 73
Nanowires, 78

Partial coherence, 134
Plasma cleaner, 68
Plasmon, 154
Probe current, 121, 146, 150, 165
Probe size, 165, 169, 171

Resolution, 5, 15, 27, 32, 33, 186
  probe size, STEM, 120, 121, 126, 135, 145
  SEM, 62, 67, 165
  TEM, 98, 103
Ronchigram, 52, 122, 126

Sample preparation (SEM), 68, 74
  Dual Beam (FIB), 75
  Broad argon ion beam, 75, 76, 85, 92
Scattering
  cross-section, 6
  factor, 6
  Rutherford, 7
  Mott, 8, 19
  Elastic, 97, 100
  Inelastic, 8, 18, 27, 97, 102, 106
SEM
  History, 2
  Instrument configuration, 11
  Secondary electrons, 17, 21
  Backscatter electrons, 18
  Contrast mechanisms, 24
  Microanalysis, 25
Semiconductors, 85
SEM column design
  Miniature, 186
  Delong/LVEM5, 187
SEM Lenses
  Electrostatic lens, 60
  Non-immersion lens, 63, 67
Signal localization, 174
Si_3N_4, 170
Specimen charging (SEM), 37, 41, 43, 45
Specimen damage, 74, 84
Specimen damage mechanisms
  Knock-on, 10, 106, 120, 131, 135, 152
  Radiolysis, 10, 106, 120, 128, 151, 154, 165, 173, 175
Spectroscopy
  EDS (EDXS), 119, 146, 150, 154
  EELS, 119, 146, 150, 154
  WDS, 91

$SrTiO_3$, 168, 175
STEM, 119, 163
STEM
  History, 2
  Instrument configuration, 11
  Imaging, 26
STEM-in-SEM, 48
  STEM converter, 49

TEM, 97
TEM
  History, 2
  Instrument configuration, 11
  Imaging, 26

Voltage contrast, 87

Wavelength, 5, 15
Wien filter, 105

X-ray, 4, 26
X-ray microanalysis
  Si(Li) detector, 196
  SDD detector, 195

Printed and bound by CPI Group (UK) Ltd, Croydon, CR0 4YY
27/09/2026
14970448-0001